Methods in Molecular Biology

Series Editor
John M. Walker
School of Life and Medical Sciences
University of Hertfordshire
Hatfield, Hertfordshire, AL10 9AB, UK

For further volumes:
http://www.springer.com/series/7651

Fast Detection of DNA Damage

Methods and Protocols

Edited by

Vladimir V. Didenko

Department of Neurosurgery and Department of Molecular and Cellular Biology, Baylor College of Medicine, Houston, Texas, USA

Humana Press

Editor
Vladimir V. Didenko
Department of Neurosurgery and Department of
Molecular and Cellular Biology
Baylor College of Medicine
Houston, TX, USA

ISSN 1064-3745 ISSN 1940-6029 (electronic)
Methods in Molecular Biology
ISBN 978-1-4939-7185-5 ISBN 978-1-4939-7187-9 (eBook)
DOI 10.1007/978-1-4939-7187-9

Library of Congress Control Number: 2017944623

Printed on acid-free paper

This Humana Press imprint is published by Springer Nature
The registered company is Springer Science+Business Media LLC
The registered company address is: 233 Spring Street, New York, NY 10013, U.S.A.

Preface

Research in DNA damage detection is a fast-paced field of study. The continuous requirement in this field for ever faster DNA damage assays is driven by the rapidly increasing numbers of samples to be analyzed and by the intensified competition between various scientific groups.

The faster assays make possible rapid high-volume and machine-based assessments. Such capability is in demand in drug design, high-throughput industry, environmental studies, and in molecular and clinical pathology research. It enables a pathology laboratory to perform rapid evaluations of DNA damage in either fixed archived tissues or unfixed clinical samples. The ubiquitous nature of DNA damage and its outcomes to a cell, such as apoptosis or necrosis, makes such assays applicable to studies across a wide range of organs and tissues. As time- and cost-saving techniques, the faster assays are useful for basic research in academia and in clinical studies of pathologies where DNA damage and apoptotic cell death play an essential role: cancer, ischemic disorders, and degenerative diseases.

To satisfy this requirement many new approaches are being developed. In addition quicker variations of the established techniques are constantly introduced. These methods use various tactics to cut down the detection time including the homogenous "closed--tube" formats with FRET reporting, new and faster labeling enzymes, quicker DNA probes, advances in the design of analytical instruments, and new ways to probe the samples, such as ultrasound scans.

Although such rapid techniques are in demand in the "research trenches," they are not covered well in the literature. This volume is the first such compendium of the time-saving techniques for detection of DNA damage and its direct physiological outcomes including apoptosis, necrosis, and phagocytic clearance.

The term "fast detection" in the book title is an umbrella designation indicating three types of time-saving assays. The assays which take around 1 h to perform comprise a group of true *express* techniques. These are followed by the *rapid* assays which take 3 h or less to complete. Next are the *accelerated* techniques. Although being lengthier than the previous two groups of methods, these represent significantly shortened, speeded up versions of routinely used techniques.

In line with this the book is divided into three parts. Part one—Express Detection—includes the fastest protocols which require less than 60 min to complete. The strategies which these express methods use to minimize the detection time include near instantaneous FRET reporting, agarose trapped cells, ultrafast labeling enzymes, as well as advantageous instrumental approaches, such as flow cytometry, and ultrasound. Most protocols in this section require 30–40 min, although several such as FRET or ultrasound assays need only 3–5 min.

Part two of the volume—Rapid Detection—describes the techniques which can be finished in 3 h or less. This part presents the isothermal "zebra tail" and RT PCR amplification assays to label DNA breaks, in addition to flow and image cytometry, and immunocytochemical detection of γ-H2AX and DNA damage. This section also contains the original Fast Micromethod procedure for genotoxicity assessments in cell suspensions and homogenized tissues.

The third part describes the accelerated detection of DNA damage and apoptosis. It contains time-saving modifications of the popular techniques. These speeded up assays mostly take 4–6 h to complete. The presented techniques include a high-throughput version of the comet assay for quicker examination of DNA damage and repair, immunofluorescence analysis of γ-H2AX foci, RAPD-PCR for the evaluation of genotoxin-induced DNA damage, and the fast-tracked ex vivo detection of γ-H2Av foci in Drosophila imaginal discs.

Overall the book presents a comprehensive collection of quick assays for the detection of nuclear and mitochondrial DNA damage and its effects in live and fixed cells and tissues, and in bacterial genomes. In addition to mammalian cells the protocols describe the use of cells from invertebrate species, such as sea mussels and Drosophila, which are convenient models for the environmental tests and toxicity studies. The volume demonstrates all levels of detection, starting from the molecular level up to the level of the entire live organism.

Such a broad methodological resource will be equally useful to both beginners in the DNA damage field and experienced researchers. Its particular usefulness is for those who perform day-to-day investigations of DNA damage and its effects, such as postdoctoral fellows, research group leaders, and academic faculty.

Researchers in molecular and cell biology, biochemistry, oncology, radiobiology, experimental and clinical pathology, toxicology, embryology, experimental pharmacology, drug design, and environmental sciences will find the book beneficial.

This book describes easily reproducible techniques requiring only a few steps to perform, and therefore, it will be useful for those who would like to gain quick access to the technical arsenal needed to study DNA damage. As such it will provide help to senior students and clinical scientists who would like to familiarize themselves with research possibilities in DNA damage detection and to explore this unique field.

I thank Candace Minchew for her highly useful and expert technical assistance. I am also grateful to Professor John Walker for his advice and encouragement throughout the review process.

Houston, TX, USA ***Vladimir V. Didenko***

Contents

Preface . . . . *v*
Contributors . . . . *ix*

PART I EXPRESS DETECTION (FROM 3 MIN TO 1 H)

1 Express FRET Labeling and Analysis of Phagocytic Clearance 3
Vladimir V. Didenko

2 Rapid Assessment of Genotoxicity by Flow Cytometric Detection of Cell Cycle Alterations 13
Nevenka Bihari

3 Ultrasound Imaging of DNA-Damage Effects in Live Cultured Cells and in Brain Tissue 23
Hadi Tadayyon, Mehrdad J. Gangeh, Roxana Vlad, Michael C. Kolios, and Gregory J. Czarnota

4 Ultrasound Imaging of Apoptosis: Spectroscopic Detection of DNA-Damage Effects In Vivo 41
Hadi Tadayyon, Mehrdad J. Gangeh, Roxana Vlad, Michael C. Kolios, and Gregory J. Czarnota

5 Fluorochrome-Labeled Inhibitors of Caspases: Expedient In Vitro and In Vivo Markers of Apoptotic Cells for Rapid Cytometric Analysis 61
Zbigniew Darzynkiewicz, Hong Zhao, H. Dorota Halicka, Piotr Pozarowski, and Brian Lee

6 The Fast-Halo Assay for the Detection of DNA Damage 75
Piero Sestili, Cinzia Calcabrini, Anna Rita Diaz, Carmela Fimognari, and Vilberto Stocchi

7 Rapid Detection of Bacterial Susceptibility or Resistance to Quinolones 95
Fátima Otero, Rebeca Santiso, María Tamayo, Germán Bou, Jaime Gosálvez, and José Luis Fernández

8 Rapid Detection of Apoptosis in Cultured Mammalian Cells 105
Igor Kudryavtsev, Maria Serebryakova, Liudmila Solovjeva, Maria Svetlova, and Denis Firsanov

9 Quick Detection of DNase II-Type Breaks in Formalin-Fixed Tissue Sections 113
Candace L. Minchew and Vladimir V. Didenko

PART II RAPID DETECTION (2–3 h)

10 Express γ-H2AX Immunocytochemical Detection of DNA Damage 123
Nate Hopp, Jodi Hagen, Birte Aggeler, and Alexander E. Kalyuzhny

11 Rapid Detection of γ-H2AX by Flow Cytometry in Cultured Mammalian Cells 129
Denis Firsanov, Liudmila Solovjeva, Olga Lublinskaya, Valeriy Zenin, Igor Kudryavtsev, Maria Serebryakova, and Maria Svetlova

12 Rapid Detection of DNA Strand Breaks in Apoptotic Cells by Flow- and Image-Cytometry 139
Hong Zhao and Zbigniew Darzynkiewicz

13 Fast Micromethod: Determination of DNA Integrity in Cell Suspensions and in Solid Tissues 151
Nevenka Bihari

14 Quantification of DNA Damage and Repair in Mitochondrial, Nuclear, and Bacterial Genomes by Real-Time PCR 159
Nadja Patenge

15 Zebra Tail Amplification: Accelerated Detection of Apoptotic Blunt-Ended DNA Breaks by In Situ Ligation 167
Vladimir V. Didenko

PART III ACCELERATED DETECTION (TIME-SAVING VERSIONS OF CONVENTIONAL TECHNIQUES)

16 Twelve-Gel Comet Assay Format for Quick Examination of DNA Damage and Repair 181
Sergey Shaposhnikov and Andrew Collins

17 Immunofluorescence Analysis of γ-H2AX Foci in Mammalian Fibroblasts at Different Phases of the Cell Cycle 187
Liudmila Solovjeva, Denis Firsanov, Nadezhda Pleskach, and Maria Svetlova

18 RAPD-PCR as a Rapid Approach for the Evaluation of Genotoxin-Induced Damage to Bacterial DNA 195
Rosanna Tofalo and Aldo Corsetti

19 Rapid Detection of γ-H2Av Foci in Ex Vivo MMS-Treated Drosophila Imaginal Discs 203
Varandt Y. Khodaverdian and Mitch McVey

Index 213

Contributors

BIRTE AGGELER • *Bio-Techne, Minneapolis, MN, USA*
NEVENKA BIHARI • *Laboratory for Marine Molecular Toxicology, Center for Marine Research, Institut Rudjer Bošković, Rovinj, Croatia*
GERMÁN BOU • *Unidad de Microbiología-INIBIC-Complexo Hospitalario Universitario A Coruña (CHUAC), A Coruña, Spain*
CINZIA CALCABRINI • *Dipartimento di Scienze Biomolecolari, Università degli Studi di Urbino "Carlo Bo", Urbino, Italy; Dipartimento di Scienze per la Qualità della Vita, Alma Mater Studiorum, Università di Bologna, Rimini, Italy*
ANDREW COLLINS • *NorGenoTech AS, Skreia, Norway; Department of Nutrition, Faculty of Medicine, University of Oslo, Oslo, Norway*
ALDO CORSETTI • *Faculty of BioScience and Technology for Food, Agriculture and Environment, University of Teramo, Teramo, Italy*
GREGORY J. CZARNOTA • *Physical Sciences, Sunnybrook Research Institute, Sunnybrook Health Sciences Centre, Toronto, ON, Canada; Department of Medical Biophysics, University of Toronto, Toronto, ON, Canada; Department of Radiation Oncology, Faculty of Medicine, University of Toronto, Toronto, ON, Canada; Department of Radiation Oncology, Sunnybrook Health Sciences Centre, Toronto, ON, Canada*
ZBIGNIEW DARZYNKIEWICZ • *Department of Pathology and Brander Cancer Research Institute, New York Medical College, Valhalla, NY, USA*
ANNA RITA DIAZ • *Dipartimento di Scienze Biomolecolari, Università degli Studi di Urbino "Carlo Bo", Urbino, Italy*
VLADIMIR V. DIDENKO • *Department of Neurosurgery and Department of Molecular and Cellular Biology, Baylor College of Medicine, Houston, TX, USA*
JOSÉ LUIS FERNÁNDEZ • *Unidad de Genética-INIBIC-Complexo Hospitalario Universitario A Coruña (CHUAC), A Coruña, Spain; Centro Oncológico de Galicia, A Coruña, Spain*
CARMELA FIMOGNARI • *Dipartimento di Scienze per la Qualità della Vita, Alma Mater Studiorum, Università di Bologna, Rimini, Italy*
DENIS FIRSANOV • *Institute of Cytology RAS, Saint-Petersburg, Russia*
MEHRDAD J. GANGEH • *Physical Sciences, Sunnybrook Research Institute, Sunnybrook Health Sciences Centre, Toronto, ON, Canada; Department of Medical Biophysics, University of Toronto, Toronto, ON, Canada; Department of Radiation Oncology, Sunnybrook Health Sciences Centre, Toronto, ON, Canada*
JAIME GOSÁLVEZ • *Unidad de Genética, Universidad Autónoma de Madrid, Madrid, Spain*
JODI HAGEN • *Bio-Techne, Minneapolis, MN, USA*
H. DOROTA HALICKA • *Department of Pathology and Brander Cancer Research Institute, New York Medical College, Valhalla, NY, USA*
NATE HOPP • *Bio-Techne, Minneapolis, MN, USA*
ALEXANDER E. KALYUZHNY • *Bio-Techne, Minneapolis, MN, USA*
VARANDT Y. KHODAVERDIAN • *Department of Biology, Tufts University, Medford, MA, USA*
MICHAEL C. KOLIOS • *Department of Physics, Ryerson University, Toronto, ON, Canada*

IGOR KUDRYAVTSEV • *Institute of Experimental Medicine, Saint-Petersburg, Russia; Far Eastern Federal University, Vladivostok, Russia*
BRIAN LEE • *Immunochemistry Technologies, Bloomington, MN, USA*
OLGA LUBLINSKAYA • *Institute of Cytology RAS, Saint-Petersburg, Russia*
MITCH MCVEY • *Department of Biology, Tufts University, Medford, MA, USA*
CANDACE L. MINCHEW • *Department of Neurosurgery, Baylor College of Medicine, Houston, TX, USA; Michael E. DeBakey Veterans Affairs Medical Center, Houston, TX, USA*
FÁTIMA OTERO • *Unidad de Genética-INIBIC-Complexo Hospitalario Universitario A Coruña (CHUAC), A Coruña, Spain; Centro Oncológico de Galicia, A Coruña, Spain*
NADJA PATENGE • *Institute of Medical Microbiology, Virology and Hygiene, University Medicine Rostock, Rostock, Germany*
NADEZHDA PLESKACH • *Institute of Cytology RAS, Saint-Petersburg, Russia*
PIOTR POZAROWSKI • *Medical University of Lublin, Lublin, Poland*
REBECA SANTISO • *Unidad de Genética-INIBIC-Complexo Hospitalario Universitario A Coruña (CHUAC), A Coruña, Spain; Centro Oncológico de Galicia, A Coruña, Spain*
MARIA SEREBRYAKOVA • *Institute of Experimental Medicine, Saint-Petersburg, Russia; State University of Information Technologies, Mechanics and Optics, Saint-Petersburg, Russia*
PIERO SESTILI • *Dipartimento di Scienze Biomolecolari, Università degli Studi di Urbino "Carlo Bo", Urbino, Italy*
SERGEY SHAPOSHNIKOV • *NorGenoTech AS, Skreia, Norway*
LIUDMILA SOLOVJEVA • *Institute of Cytology RAS, Saint-Petersburg, Russia*
VILBERTO STOCCHI • *Dipartimento di Scienze Biomolecolari, Università degli Studi di Urbino "Carlo Bo", Urbino, Italy*
MARIA SVETLOVA • *Institute of Cytology RAS, Saint-Petersburg, Russia*
HADI TADAYYON • *Physical Sciences, Sunnybrook Research Institute, Sunnybrook Health Sciences Centre, Toronto, ON, Canada; Department of Medical Biophysics, University of Toronto, Toronto, ON, Canada; Department of Radiation Oncology, Sunnybrook Health Sciences Centre, Toronto, ON, Canada*
MARÍA TAMAYO • *Unidad de Genética-INIBIC-Complexo Hospitalario Universitario A Coruña (CHUAC), A Coruña, Spain; Centro Oncológico de Galicia, A Coruña, Spain*
ROSANNA TOFALO • *Faculty of BioScience and Technology for Food, Agriculture and Environment, University of Teramo, Teramo, Italy*
ROXANA VLAD • *Department of Radiation Oncology, Juravinski Cancer Centre, Hamilton Health Sciences, Hamilton, ON, Canada*
VALERIY ZENIN • *Institute of Cytology RAS, Saint-Petersburg, Russia*
HONG ZHAO • *Department of Pathology and Brander Cancer Research Institute, New York Medical College, Valhalla, NY, USA*

Part I

Express Detection (from 3 min to 1 h)

Chapter 1

Express FRET Labeling and Analysis of Phagocytic Clearance

Vladimir V. Didenko

Abstract

Lysosomal DNase II in phagocytic digestion produces DNA ends with $3'PO_4/5'OH$, which differ from those created in apoptotic DNA fragmentation, and can be used to label phagocytic clearance of cell death. Here, we describe the use of these specific DNA ends as selective markers of phagocytic reaction in cell suspensions. The approach does not require cell fixation. It selectively labels blunt-ended DNA breaks with terminal 5′OH. The detection is performed by ultra-fast FRET probes in a single step, closed-tube procedure. It takes 3 min and is signaled by fluorescence. The full step-by-step protocol is presented as well as instructions on analysis and representation of the results.

The described DNA-end-based phagocytosis marker and the new rapid FRET assay can be useful in studies of phagocytosis, apoptosis and in immune system assessments.

Key words FRET probes, Nanoblinkers, Phagocytic digestion of DNA, 5′OH DNA breaks labeling, Phagocytosis, Phagolysosomes, Apoptotic cell clearance, Express detection of DNase II cleavage, 5′OH DNA probes

1 Introduction

Apoptotic fragmentation and phagocytic digestion of DNA are the only two processes in cells which can specifically generate large numbers of blunt-ended DNA breaks. However, these two pools of breaks differ principally in the distribution of their functional end-groups. Apoptotic nucleases produce DNA breaks with $3'OH/5'PO_4$, while lysosomal DNase II in phagocytic digestion makes the inverted $3'PO_4/5'OH$ configuration [1, 2].

In another report in this volume, we described an approach for fast and specific labeling of such DNase II-type breaks in situ, in fixed tissue sections [3]. The procedure permitted selective visualization of active phagocytes performing clearance of apoptotic cells. On a subcellular level, the in situ assay labeled phagolysosomes in the cytoplasm of the cells that engulfed apoptotic nuclei. These cellular organelles are responsible for the actual

Vladimir V. Didenko (ed.), *Fast Detection of DNA Damage: Methods and Protocols*, Methods in Molecular Biology, vol. 1644, DOI 10.1007/978-1-4939-7187-9_1, © Springer Science+Business Media LLC 2017

execution of phagocytic degradation of apoptotic cell corpses, and are the final sites where engulfed DNA material is destroyed by DNase II [4, 5].

Here, we expand the labeling of DNase II-type cleavage to live cell suspensions, such as cultured non-adherent cells. The technique can also be adapted to noncultured cells, such as those obtained in blood tests, for example peripheral blood mononuclear cells. Moreover, the procedure can be used with adherent cells after they are put in suspension.

The approach detects blunt-ended DNA breaks with terminal 5′OH, a general marker of phagocytic DNA digestion [4, 6]. Therefore, it can label all the various cell types participating in phagocytic clearance.

By being a FRET-based technology, this assay differs significantly from the previously described in situ labeling techniques [3, 4]. This is because FRET probes, which can work in a homogenous cell culture assay, have different requirements as compared to the in situ probes for tissue sections.

In a homogenous closed-tube format, the unreacted probe cannot be washed out from the reaction mix making it difficult to determine the amount of probe which actually reacted with its target. To overcome this limitation we developed a new type of FRET probes termed nanoblinkers, which do not require washing steps as they indicate detection of their targets by changing their fluorescence color. Specifically, the FRET probes described in this presentation signal the detection of specific DNA breaks by changing their fluorescence emission maxima from 580 to 525 nm (i.e., from red to green).

The general structure of the nanoblinker FRET probe and its mode of action are presented in Fig. 1. The figure shows that the probe is a dual-hairpin oligonucleotide with two fluorophores (FRET donor and acceptor) positioned in close proximity to enable FRET (*see* **Note 1**).

The actual oligonucleotide probe that we use in the described protocol is a 38-mer dual-hairpin labeled by FAM-TAMRA. It contains CCCTT3′ sequence recognized by the TOPO enzyme. The enzyme specifically attaches to this site and makes a single cut at its 3′ end (CCCTT ↓3′). This breaks the dual-hairpin into two separate hairpins, one carrying the enzyme and another enzyme-free (Fig. 1). The repetitive cleavage-religation permits separation and random re-association of the hairpins. This oscillating system is a specific detector of 5′OH DNA breaks because the length of the donor fluorescence phase, when the hairpins are separated, radically increases in their presence. A blunt-ended 5′ OH DNA break represents the selective alternative target for the acceptor-carrying hairpin with bound TOPO, which can ligate to it instead

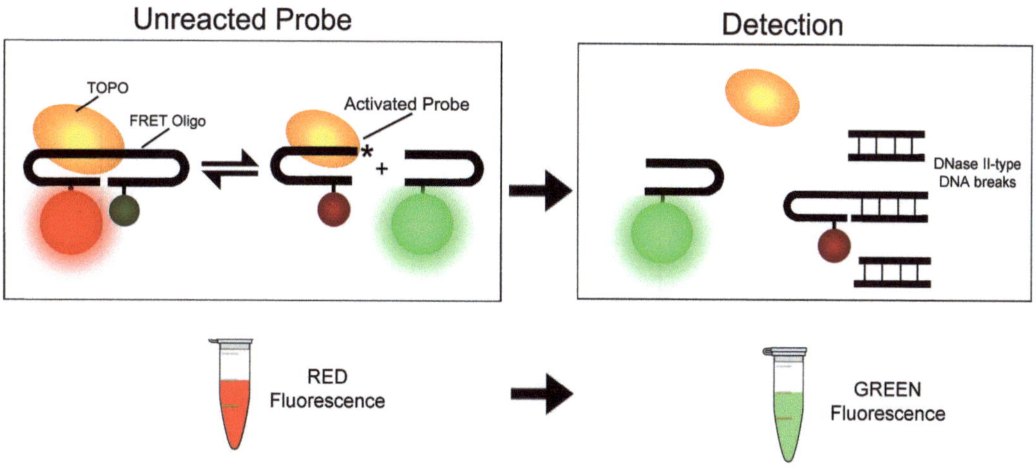

Fig. 1 Nanoblinker FRET assay for detection of DNase II breaks in cell suspensions. In the dual hairpin FRET Oligo the *green* DONOR fluorescence is suppressed due to FRET-based transfer of energy to the *red* ACCEPTOR (FAM and TAMRA fluorophores in the actual probe). Vaccinia topoisomerase I (TOPO) binds to the FRET Oligo and cleaves it in the center. Two hairpins perpetually ligate back and then recleave, producing alternating DONOR–ACCEPTOR fluorescence. In this equilibrium the ligated state is predominant and the main fluorescence is from the ACCEPTOR [7]. When DNase II breaks are present, the ACCEPTOR part ligates to them instead. This permanently stops FRET, eliminates *red* ACCEPTOR fluorescence, and activates *green* DONOR fluorophore in its place. The *color change* indicates detection

of the donor-labeled hairpin [4, 8]. This interrupts FRET, shifts fluorescence emission from acceptor to donor (i.e., probe fluorescence changes from red to green), thus signaling detection of DNA damage (Fig. 2).

The nanoblinker FRET probe can uncover DNA damage in the homogenous assay format without the need to remove the unreacted probe. The assay is sensitive and in tests could distinguish between apoptotic and necrotic cell engulfment, based on the intensity of phagocytic clearance [7] (*see* **Note 3**).

Here, we present the complete protocol of this ultra-fast FRET technique applicable for detection of DNase II-type breaks in suspensions of unfixed, live cells. With its labeling step requiring only 3 min and no need for cell fixation, the assay provides the most rapid evaluation of specific DNA breaks. The nanoblinker FRET probe reacts exclusively with 5′OH blunt-ended DNA breaks. The detection is not affected by the presence of other types of DNA damage, beyond the target breaks of DNase II-type [7].

The described protocol was extensively used in studies of phagocytic J774A.1 cells clearing apoptotic and necrotic U87 cells [7]. The assay does not change when any other cell types are used, but in some cases it might need an adjustment.

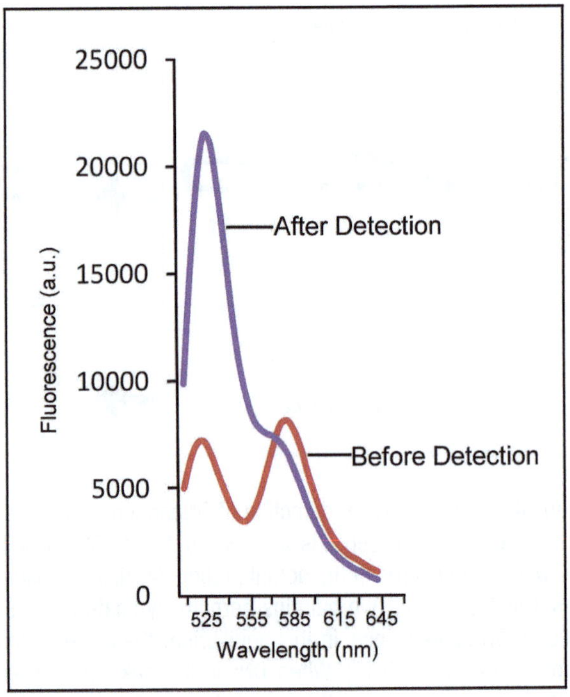

Fig. 2 Nanoblinker FRET probes detect 5′OH DNA ends. Emission spectra of cycling nanoblinkers before and after the addition of 5′OH DNA ends (Test Oligo). Before the reaction the emission maxima correspond to fluorescence of both DONOR (FAM—525 nm) and ACCEPTOR (TAMRA—580 nm). The detection reaction increases fluorescence of DONOR and decreases that of ACCEPTOR. $\lambda_{Excitation}$–488 nm. (*see* **Note 2** for full test details)

2 Materials

1. Phagocytic J774A.1 cells from ATCC (TIB-67).
2. U87 cells from ATCC (HTB-14).
3. FRET Oligo—a 38-mer oligonucleotide dual-labeled with FAM and TAMRA:

   5′-AAGGGT(**TAMRA**)CCTGCTGCAGGACCCTTAACGC ATTATGCGT(**FAM**)T-3′.
4. Test Oligo—a 21-mer hairpin used to emulate the blunt-ended DNase II breaks in positive control solution tests: 5′-GCGCTAGACCTGGTCTAGCGC-3′.

   All oligonucleotides synthesized and PAGE purified by IDT, Coralville, IA. On receipt dilute oligonucleotides with bidistilled water to 100 pmol/μL concentration and store at −20 °C, protected from light.

5. Vaccinia DNA topoisomerase I—3000 U/μL (Millipore) (*see* **Note 4**).
6. 100 mM Tris–HCl, pH 7.4.
7. Scanning spectrofluorometer (Tecan Safire 2 or similar).
8. APO HTS 3/7 Caspase Detection kit (Cell Technology, Inc.). This kit is used for several purposes in the assay. First, it helps to confirm the apoptotic morphology of the cells. Second, it provides the cell lysis buffer that is also used for the disruption of cellular membranes to make the cellular contents available for the analysis by the nanoblinkers (*see* **Note 5**).

3 Method

3.1 Induction of Apoptosis, Necrosis, and Phagocytosis

1. Induce apoptosis in cultured U87 cells: incubate at 42 °C for 30 min, then return cells to 37 °C for 18 h.
   Induce necrosis in cultured U87 cells by incubating cells at 65 °C for 10 min. Then return cells to 37 °C for 18 h.
2. Verify apoptosis and necrosis by using the APO HTS 3/7 Caspase Detection kit and morphologically by fluorescence microscopy using DAPI staining (*see* **Note 6**).
3. Induce phagocytosis by combining treated U87 cultures (200 cells/well) with cultured J774A.1 macrophages (20 cells/well) for 18 h at 37 °C, 5% CO_2 (*see* **Note 7**).

3.2 Phagocytosis Assessment with Nanoblinker FRET Probes

1. After the18 h-long incubation place the combined macrophage and U87 cell cultures (20 macrophages/200 U87 cells) and all negative controls in the cell lysis buffer and vortex for 30 s to rupture cellular membranes (*see* **Note 5**).
2. Aliquot all samples into a 96-well plate.
3. Assemble nanoblinkers in 100 mM Tris–HCl, pH 7.4 by combining 50 pmol FRET Oligo and 50 pmol vaccinia topoisomerase I. Incubate for 1 min at room temperature. Add 20 μL of prepared nanoblinkers (50 pmol total) to 80 μL (20 macrophages/200 U87 cells) of ruptured cells diluted in 100 mM Tris-HCl, pH 7.4 in a 96-well plate. The final concentration of nanoblinkers in the wells will be 500 fmol/μL in 100 μL total volume (*see* **Notes 2** and 7).
4. Incubate for 3 min at room temperature, protected from light. For the incubation the plate can be placed in the spectrofluorometer ready for the measurements (*see* **Note 8**).
5. Immediately after the incubation simultaneously measure both donor (525 nm) and acceptor (580 nm) emissions using 488 nm excitation directly in the 96-well plate. Use the obtained values for the assessment of DNase II-type breaks in

the samples (*see* **Note 9**). Compare experimental results with controls present in the same 96-well plate (i.e., nanoblinker probe signal without 5′OH DNA breaks) (*see* **Note 7**).

4 Notes

1. In FRET the rate of energy transfer is inversely proportional to the sixth power of the distance between the donor and acceptor fluorophores. Therefore, the efficiency of the transfer very rapidly declines with distance. For each specific donor-acceptor pair a distance (in Å) at which energy transfer is 50% efficient is called their Förster radius ($\mathbf{R_0}$) [9]. The efficiency of FRET swiftly drops to zero at distances larger than the Förster radius. The Förster radii are experimentally determined for each pair. For FAM and TAMRA pair $\mathbf{R_0} = 55$ Å [10].

   In the nanoblinker FRET Oligo FAM and TAMRA are at the distance of 23.8 Å [7]. Therefore, FRET efficiency ($\mathbf{E_{FRET}}$) in this case is:

$$\mathbf{E_{FRET}} = \mathbf{1} / \left[\mathbf{1} + \left(\mathbf{r} / \mathbf{R_0} \right)^{\mathbf{6}} \right] = 1/1.006565769 = 0.993477$$

   In other words, FRET efficiency between FAM and TAMRA at this distance is 99.35%. Therefore in the uncleaved oligo the acceptor (TAMRA—red) is radiative, whereas the donor (FAM—green) is completely quenched. We investigated the nanoblinker FRET system in more detail elsewhere [7].

2. We recommend testing the nanoblinker system before performing any experiments with cells. Such controls can confirm the assay sensitivity. For these perform detection by using the exogenous 5′OH blunt-ended DNA in solution.

   For the tests assemble the nanoblinker FRET probes in 100 mM Tris–HCl, pH 7.4 by combining 10 pmol (2 pmol/μL final concentration) FRET Oligo and 10 pmol (2 pmol/μL final concentration) vaccinia topoisomerase I (TOPO). To the reaction mix add 100 pmol of Test Oligo (20 pmol/μL final concentration) in 100 mM Tris–HCl, pH 7.4. Incubate the reaction for 1 min at room temperature before measurements. Scan samples using a spectrofluorometer at 488 nm excitation, or measure fluorescence of the donor (FAM) at 488 nm excitation and 525 nm emission wavelengths. The increase in 525 nm fluorescence indicates a positive reaction (*see* **Note 9** for more options).

   Occasionally additional controls might be needed to exclude possible contamination of vaccinia topoisomerase preparations with nucleases. In such cases, the pretreatment of control sections with TOPO for 2 h at 37 ° C is recommended followed by detection of DNA breaks.

3. The sensitivity of the assay depends on the biological model used, the labeling conditions, and the equipment for signal registration. In the biochemical tests using a Tecan Safire 2 scanning spectrofluorometer, the nanoblinker assay dependably detected 100 pmol of 5′OH DNA ends, represented by the Test Oligo (20 pmol/mL final concentration), after 30 s of incubation. In cell culture conditions after 3 min of incubation the assay was capable of detecting phagocytosis and reliably distinguished between apoptotic and necrotic cell engulfment based on the intensity of phagocytic activity of 20 phagocytes digesting 200 apoptotic or necrotic cells [7]. When used with fluorescence microscopy as in situ probes, the nanoblinkers were applied to the cells grown on glass chamber slides. In these conditions the assay after 15 min of incubation fluorescently tagged individual phagolysosomes in macrophages digesting apoptotic and necrotic cells and enabled the subcellular level of analysis [7].
4. Active vaccinia topoisomerase I, which works well with the described assay, can be purchased from Millipore, sold as a part of the ApopTag® ISOL Dual Fluorescence Apoptosis Detection Kit. In most experiments we used the highly concentrated preparation of this enzyme obtained from Vivid Technologies (Houston, TX).
5. Cell lysis buffer is a proprietary phosphate-based buffer with pH ~ 7 supplied as part of the APO HTS 3/7 Caspase Detection kit from Cell Technology. Other approaches can be used to disrupt cells instead of this buffer, such as hypo-osmotic shock or freezing-thawing of cells. Prior to the application these approaches should be experimentally tested with the specific cell samples that will be investigated by the nanoblinker probes.
6. APO HTS 3/7 Caspase Detection kit from Cell Technology is the homogenous fluorimetric assay for active caspase 3/7, using $\lambda_{excitation}$ = 488 nm, $\lambda_{emission}$ = 525 nm.
7. As negative controls for the18 h-long phagocytic reaction use cells without 5′OH DNA breaks, such as normal macrophages, apoptotic U87 cells, and necrotic U87 cells. All negative controls cells (apoptotic, necrotic, and normal) should be incubated at 37 C for 18 h in parallel with the other (phagocytic) series of the experiment.
8. Strict 3 min incubation time was verified as sufficient for the sensitive measurement of specific DNA breaks in cultured cells [7]. However, the results also confirmed that the detection capability of the nanoblinker system was maintained for at least

15 min post assembly [7]. Therefore, it is highly likely that a different incubation interval within this time range can be used, if needed, but should be experimentally tested beforehand. In any case, the same exact incubation time should be used for all series.

9. FRET cessation in the TOPO probe signals detection of DNA breaks. Therefore, both the peak fluorescence of the donor at 525 nm and the ratio of donor and acceptor emissions at the excitation wavelength of the donor ($\mathbf{E}_{D/A} = \mathbf{E}_{D\ 525\ nm}/\mathbf{E}_{A\ 580\ nm}$) can be used for the assessment of samples [7, 11, 12].

   However, the use of the $\mathbf{E}_{D/A}$ ratio provides a more sensitive measure because the disruption of FRET simultaneously increases the donor emission and diminishes the signal from the acceptor. Combined these two effects drive the $\mathbf{E}_{D/A}$ ratio up [11, 12].

Acknowledgment

I am grateful to Candace Minchew for her outstanding technical assistance.

This research was supported by grant R01 NS082553 from the National Institute of Neurological Disorders and Stroke, National Institutes of Health and by grants R21 CA178965 from the National Cancer Institute, National Institutes of Health and R21 AR066931 National Institute of Arthritis and Musculoskeletal and Skin Diseases, National Institutes of Health.

References

1. Didenko VV, Minchew CL, Shuman S, Baskin DS (2004) Semi-artificial fluorescent molecular machine for DNA damage detection. Nano Lett (12):2461–2466
2. Didenko VV (2011) 5′OH DNA breaks in apoptosis and their labeling by topoisomerase-based approach. Methods Mol Biol 682:77–87
3. Minchew CL, Didenko VV (2017) Quick detection of DNase II-type breaks in formalin-fixed tissue sections. Methods Mol Biol 1644:113–119
4. Minchew CL, Didenko VV (2011) Fluorescent probes detecting the phagocytic phase of apoptosis: enzyme-substrate complexes of topoisomerase and DNA. Molecules 16:4599–4614
5. Nagata S, Nagase H, Kawane K et al (2003) Degradation of chromosomal DNA during apoptosis. Cell Death Differ 10:108–116. doi:10.1038/sj.cdd.4401161
6. Minchew CL, Didenko VV (2014) Assessing phagocytic clearance of cell death in experimental stroke by ligatable fluorescent probes. J Vis Exp 87:e51261. doi:10.3791/51261
7. Minchew CL, Didenko VV (2014) Nanoblinker: Brownian motion powered bio-nanomachine for FRET detection of phagocytic phase of apoptosis. PLoS One 9:e108734. doi:10.1371/journal.pone.0108734
8. Minchew CL, Didenko VV (2012) In vitro assembly of semi-artificial molecular machine and its use for detection of DNA damage. J Vis Exp 59:e3628
9. Didenko VV (2001) DNA probes using fluorescence resonance energy transfer (FRET): designs and applications. BioTechniques 31:1106–1121
10. Haugland RP (2010) Handbook of fluorescent probes and research products, 11th edn. Life Technologies, Grand Island

11. Miyawaki A, Tsien RY (2000) Monitoring protein confirmations and interactions by fluorescence energy transfer between mutants of green fluorescent protein. Methods Enzymol 327:472–500

12. Hailey DW, Davis TN, Muller EG (2002) Fluorescence resonance energy transfer using color variants of green fluorescent protein. Methods Enzymol 351:34–49

Chapter 2

Rapid Assessment of Genotoxicity by Flow Cytometric Detection of Cell Cycle Alterations

Nevenka Bihari

Abstract

Flow cytometry is a convenient method for the determination of genotoxic effects of environmental pollution and can reveal genotoxic compounds in unknown environmental mixtures. It is especially suitable for the analyses of large numbers of samples during monitoring programs. The speed of detection is one of the advantages of this technique which permits the acquisition of 10^4–10^5 cells per sample in 5 min. This method can rapidly detect cell cycle alterations resulting from DNA damage. The outcome of such an analysis is a diagram of DNA content across the cell cycle which indicates cell proliferation, G_2 arrests, G_1 delays, apoptosis, and ploidy.

Here, we present the flow cytometric procedure for rapid assessment of genotoxicity via detection of cell cycle alterations. The described protocol simplifies the analysis of genotoxic effects in marine environments and is suitable for monitoring purposes. It uses marine mussel cells in the analysis and can be adapted to investigations on a broad range of marine invertebrates.

Key words Genotoxicity assessment, Cell cycle alterations, DNA content, Flow cytometry, Marine mussel, *Mytilus galloprovincialis*

1 Introduction

Analysis of DNA content and cell cycle alterations by flow cytometry is widely used in detection of genotoxic and/or xenobiotic effects in different cell types from vertebrates and invertebrates. The DNA content of cells is assessed with the help of the stoichiometrically binding DNA dyes, so that the emitted fluorescence is proportional to the DNA amount. The most widely used dye is 4′ 6-diamidino-2-phenylindole (DAPI). DAPI has a strong A-T base preference and yields DNA histograms with coefficients of variation (CVs) lower than that obtained using other dyes. Additionally, staining with DAPI is less affected by the state of chromatin condensation compared to other DNA dyes.

Flow cytometric DNA analysis is performed to determine the percentage of cells in each phase of the cell cycle and to evaluate

Vladimir V. Didenko (ed.), *Fast Detection of DNA Damage: Methods and Protocols*, Methods in Molecular Biology, vol. 1644, DOI 10.1007/978-1-4939-7187-9_2, © Springer Science+Business Media LLC 2017

cell cycle alterations in cellular populations. As a rule, all normal diploid cells (non-replicating G_0 cells and cells in G_1 phase of cell cycle) in the same eukaryotic organism should have the same DNA content. This quantity is usually expressed as 2n. DNA synthesis during the S phase of the cell cycle results in an increase in cellular DNA content, which reaches 4n at the end of S phase and remains at this value during the G_2 phase and during mitosis (M phase). After the completion of mitosis the original cell is replaced by two daughter cells, each with DNA content of 2n. A typical DNA content distribution across a cell cycle obtained by flow cytometry is shown in Fig. 1. The analyzed cells fall into three categories: (1) Cells in G_0 or G_1 phase, i.e., with an unreplicated complement of DNA; (2) Cells in G_2 or M phase with a fully replicated complement of DNA (twice the G_1 DNA content); (3) Cells in S phase, with an intermediate amount of DNA.

The distribution of cells in Fig. 1 indicates that there are more cells in the G_0/G_1 phase than in the G_2/M phase showing that G_0/G_1 is longer than G_2/M in this population. The DNA content distribution always exhibits some variance in the G_0/G_1 peak, which may be due to staining procedures, to instrumental errors, and/or to cell-to-cell differences in DNA content.

The obtained DNA content distribution is analyzed automatically by the software package supplied with the flow cytometer. The analysis includes the calculation of CV, i.e., the standard deviation of the distribution for the diploid peak divided by the peak mean, and the percentage of cells in G_0/G_1, S, and G_2/M phases during the cell cycle. Higher CVs can often result from chromosomal aberrations caused by clastogenic agents—mutagens inducing disruption or breakages of chromosomes [1]. A larger CV may also be due to the partial inclusion of an aneuploidy peak in the

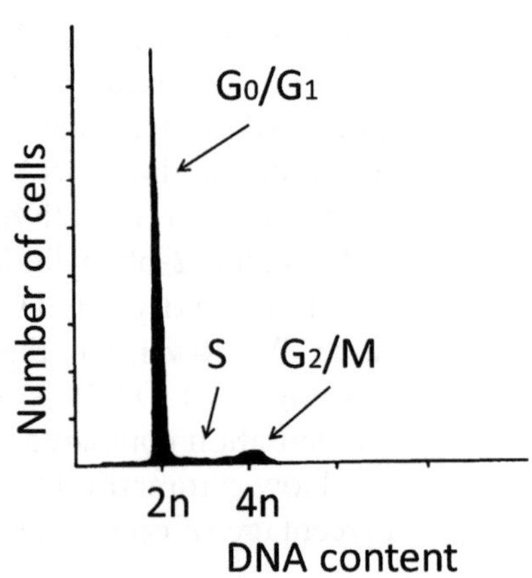

Fig. 1 Typical DNA content histogram obtained with flow cytometry

diploid peak [2]. There is an established link between the increase in CV and both chemical [3] and radiation exposure [4]. Increases in CVs have been reported for erythrocytes of turtles exposed to low-level radiations in effluent ponds of nuclear power plants [4], for blood samples of green frogs exposed to pesticides [5], for vertebrates exposed to radionuclides and other mutagenic chemicals [4, 6], for blood in fish from Chernobyl-contaminated ponds [7], and, more recently, for clams collected at polluted sites of Sagueny Fjord, Canada [8].

Flow cytometry is useful in determining the pollutant-induced genetic damage in marine invertebrate populations. Flow cytometric analysis was successfully used in studies of DNA alterations in different organs of the marine mussel *Mytilus galloprovincialis* [9]. Mussels of the genus *Mytilus* are among the most common of the marine molluscs, constituting an important element of both the ecology of coastal waters and the economy as food and fouling organisms. They accumulate various contaminants from the surrounding water and serve as sensitive bioindicators of coastal water quality. Because hemolymph and gills of marine mussels respond first to genotoxic agents, they are used for the monitoring of environmental conditions. Cell cycle alterations in the hemolymph DNA were reported for mussels collected at sampling sites that were under the influence of anthropogenic loads [10] and in mussels treated in vivo with the herbicide 2,4-dichlorophenoxy acetic acid [11]. Induction of internucleosomal DNA fragmentation, i.e., apoptosis, was detected in gill tissue from mussels treated with tri-*n*-buthyltin chloride [12].Thus, the flow cytometric analysis of DNA content in marine mussels can be used as a pollution test in the ecosystem survival studies of polluted areas.

This chapter describes the flow cytometric procedure for the rapid assessment of genotoxicity via detection of DNA content and cell cycle alterations. The sample preparation is shortened and does not require the hemocyte isolation step. The described protocol simplifies the analysis of genotoxic effects in marine environments and is suitable for monitoring studies. It uses mussel *Mytilus galloprovincialis* hemolymph and gills cells in the analysis and can be adapted to investigations on a broad range of marine invertebrates.

The step-by-step description includes the instrument setup and standardization, sample preparation, DNA content measurement, data analysis, and reporting. We also provide several examples of altered DNA content histograms obtained by flow cytometric analysis. The described protocol requires up to 5 min for the acquisition of 10^4–10^5 cells (one sample) and allows analysis of more than 50 samples per day. The procedure gives information about several cell cycle alterations in the analyzed samples.

2 Materials

1. Standards: trout and chicken erythrocyte nuclei as indicators of DNA ploidy (commercially available).
2. 4′ 6-diamidino-2-phenylindole dihydrochloride (DAPI) staining solution for nonfixed cells (Partec GmbH, Germany).
3. Flow Cytometer (Partec PAS II or Partec PAS III that we used, or other available models).
4. Hemolymph probe (*see* **Note 1**): add 100 μL of hemolymph (1–2 × 10^6 hemocytes per mL of hemolymph) to the 1 mL DAPI staining solution containing 10% of dimethyl sulfoxide (DMSO) and analyze immediately.
5. Gills probe: 1 mg of gills (about 1 × 10^6 cells) gently resuspend in DAPI staining solution; filter through 30 μm filter and analyze immediately.

3 Method

3.1 Instrument Setup

Instrument setup varies with manufacturer. However, there are some general principles to observe.

1. Select the LIN channel that is most appropriate for the DNA probe.
2. Set the trigger on the channel detecting the DNA probe.
3. Select parameters to enable doublet discrimination (*see* **Note 2**).
4. Set a gate to exclude doublets and apply it to the histogram that will display the DNA profile.
5. Make sure the sheath tank is full, as it may help with stability.
6. Make sure the cytometer is clean. Stream disruption will increase the CV (*see* **Note 3**).
7. Set a low flow rate and dilute cells to a concentration that is appropriate for the DNA probe solution.
8. Make sure the instrument has been optimized by running routine calibration particles.

3.2 Standardization

1. Check the performance of the instrument on a daily basis using commercially available fluorescent beads of known CV or DNA standards (*see* **Note 4**). Any perturbation of the sample stream in the cytometer will increase the CV and for this reason the concentration of cells or nuclei should be kept high (between 5 × 10^5 and 2 × 10^6/mL) and the flow rate low.
2. Check that there is not a partial blockage of the flow cell. Report the CV of the main G_0G_1 peak. Generally, less than 3 is good; greater than 8 is poor.

3.3 Sample Preparation

1. Take the hemolymph from the adductor muscle of the mussel (*see* **Note 5**) and prepare the hemolymph probe (*see* **Note 6**).
2. Dissect the gills from the mussel and prepare the gills probe.

3.4 DNA Content Measurement

1. Analyze the hemolymph or gill probe with the flow cytometer: flow rate 200–400 cells/s, excitation of 100 W mercury lamp, emission 455 nm for the DAPI signal or excitation with UV laser at 359 nm, and emission at 461 nm. Analyze 2×10^4 hemocytes or gill cells from each sample. First, make the acquisition of the control sample (stained hemolymph or gills of untreated healthy specimen) and then of the investigated sample.
2. Repeat measurements in triplicates and rerun after adding 10 μL of mussel sperm internal standard (*see* **Note 7**) in order to calculate the DNA index (DI). DI is generally defined as the ratio between the DNA content of a tumor cell and that of a normal diploid cell. In the mussel protocol the DI value is calculated as the ratio between the position of the diploid peak of the investigated cells and that of the mussel sperm DNA (internal standard), as discussed in Subheading 3.5. Data analysis and reporting.
3. Finally, rerun control samples after each individual series (*see* **Note 8**). Variability due to differences in sample preparations, staining procedures, condition of the mercury lamp, and adjustments of the flow cytometer optical system should be fairly constant (small samples CV standard deviation) (*see* **Note 9**).

3.5 Data Analysis and Reporting

1. Calculate the DI value as the ratio between the position of the G_0/G_1 peak of the affected cell line or tissue and that of normal diploid cells (control), in which case it should be 1 ± 0.05 [2].

   The flow cytometer provides a histogram of the DNA content distribution across a cell cycle. Altered DNA content distribution is immediately observable from the histogram. Furthermore, CVs as well as the percentage of cells in each phase of the cell cycle are already provided by the instrument's software package, while the DI value should be calculated. To analyze if the CV reflects the affected cells compare only CVs of normal and symmetrical DNA content histograms (Table 1).
2. In the case of mussels calculate the DI value as the ratio between the position of the diploid peak of the investigated cells and that of the mussel sperm DNA (internal standard) (*see* **Notes 10** and **11**).

Table 1
Coefficient of variations for G_0/G_1 peak of mussel hemolymph

Mussels	N	CV	Confidence interval
Maricultured	30	4.59	4.22–4.77
Injected with DMSO as a vehicle for different contaminants	15	4.85	4.63–5.07
Collected along Adriatic coast	20	5.11	4.95–5.28
Collected at site under the anthropogenic load	10	6.87	6.04–7.42

4 Notes

1. It is not necessary to isolate and use a particular hemocyte type from mussel hemolymph since hemolymph cell kinetics is uniform although it contains different hemocytes. The reason is low DNA content variation in different hemocyte types, not detectable by flow cytometry.
2. Two nuclei or cells in G_1 of the cell cycle stuck together will have the same DNA content as a single cell in G_2 and the two should be distinguished if the DNA histogram is to reflect accurately the state of the cell cycle. Doublets will give a wider signal than single cells while cells in G2/M will give a higher peak signal compared to two clumped cells in G1 (doublet) but a narrower width. Some instruments are designed to display width against area; other models display peak height against area.
3. The CV as a measure of instrument precision and the peak channel number for a standard set of conditions (laser power, photomultiplier voltage, and gain) should be recorded. Predetermined limit for CV is 2% for calibration particles or stained nuclei since DNA content is so precisely regulated so as to vary by less than 2% from cell to cell in homogeneous, nondividing populations. Restore the instrument's performance if these fall outside the predetermined limit.
4. Perform the standardization of the instrument and the acquisition of a control sample to distinguish between effects of the instrument setup and effects of genotoxins in the examined cells. During the standardization the following criteria for mussel tissues must be achieved:
   - Acquisition of at least 20,000 nuclei.
   - Low flow rate (100–200 cells/s) for narrow CV.

- G_0/G_1 peak on a suitable scale in a known channel.
- CV of the G_0/G_1 peak 5 ± 0.5%.
- S phase containing at least 200 cells.
- Data containing less than 3% debris.
- G_0/G1 to G2 M ratio between 1.95 and 2.05.

If the above is achieved for control mussel tissues any change in the DNA content histogram, increase in CV, or change of DI value of the analyzed sample could be attributed to the alteration of the cycle caused by genotoxins.

5. Open the mussel valves carefully and just enough to insert the needle of a syringe without damaging the adductor muscle. Slightly insert the needle into the adductor muscle and withdraw 100 μL of hemolymph. It is possible to withdraw up to 600 μL of hemolymph from a mussel 4–5 cm in length. Be careful not to tear the adductor during the process. If necessary, check the hemolymph content for the presence of hemocytes using a microscope.
6. Prepare the hemolymph probe immediately at the site of mussel collection and freeze it in liquid nitrogen for transportation to the laboratory. Store at −80 °C up to 1 year.
7. As instrument calibration standard use commercially available trout erythrocytes. As an internal fluorescence standard use freshly isolated and fixed mussel sperm, obtained as described in [13]. Briefly, the mussel sperm is resuspended in a buffer-fluorochrome solution. The latter is prepared by adding 10% DMSO to combined solutions A (4 parts) and B (1 part), just prior to use. Solution A: 0.85% w/v NaCl, 0.1 M Tris, 1 mM $CaCl_2$, 0.5 mM $MgCl_2$, 0.2% w/v bovine serum albumin, adjusted to pH 8.0; Solution B: 106 mM $MgCl_2$, 0.6% w/v Triton X-100, 50 μg/mL DAPI. Samples should be kept frozen at −70 °C until use.
8. Rerunning the control sample is a necessary step in order to rinse and check the performance of the instrument between sequentially analyzed samples.
9. CVs strongly depend on the type of flow cytometer used for measurements, e.g., for healthy mussel hemolymph the CV of the diploid peak calculated by the software package has been 3.5 ± 0.5 (n = 60) and 4.5 ± 0.5 (n = 57) for two instrument types we tested (Partec PAS II and Partec PAS III), respectively.
10. For normal, healthy mussel specimens, DI calculated relative to mussel sperm is 1.8–2.0.
11. Examples of altered DNA content histograms obtained by flow cytometric analysis are shown in Figs. 2 and 3. These demon-

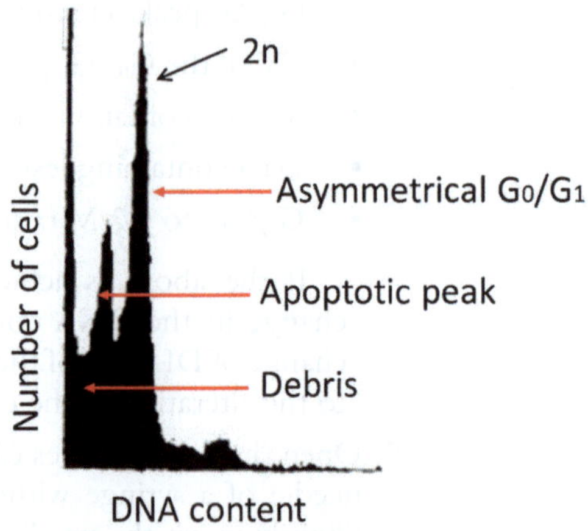

Fig. 2 DNA content histogram of apoptotic mussel hemolymph cells

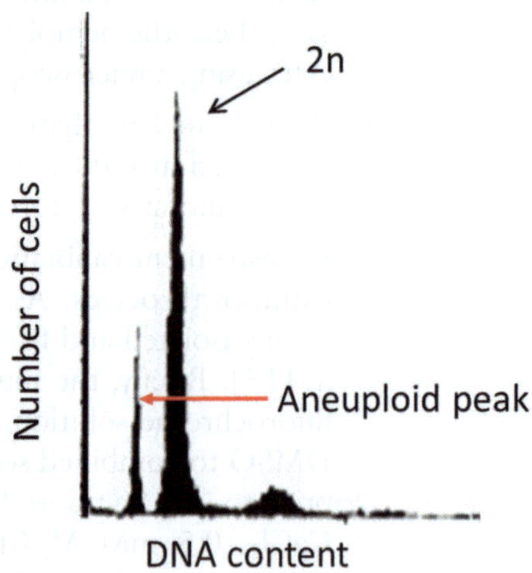

Fig. 3 DNA content histogram of aneuploid mussel hemolymph cells

strate two examples of altered DNA profiles obtained for hemolymph of several mussels collected along the Adriatic coast: apoptotic DNA content histogram (Fig. 2) and DNA content histogram that reveal aneuploidy (Fig. 3). Apoptosis is characterized by (1) An asymmetrical G_0/G_1 peak, (2) A decrease of the G_2/M peak concomitant with the appearance of cells characterized by low DNA content, i.e., below that of the G_0/G_1 peak (so-called apoptotic peak) and (3) Increase in the number of dead cells appearing as debris at the far left side of the DNA content distribution. DNA content distribution describing aneuploidy shows an additional peak on the left side of the diploid peak arising from chromosome loss [14]. Aneuploidy is also specified by the decrease in DI values [15].

For mussel hemolymph, aneuploidy with only hypodiploid DNA content, DI lower than 0.9 was detected in 30.8% of the 146 investigated mussels collected along the Adriatic coast [9].

Acknowledgment

Special thanks are given to Iris Batel for her corrections, suggestions, and comments that substantially improved this chapter.

References

1. Galloway SM (1994) Chromosome aberrations induced in vitro: mechanisms, delayed expression and intriguing questions. Environ Mol Mutagen 23(Suppl 24):44–53
2. Sun T (1993) Color atlas—text of flow cytometric analysis of hematologic neoplasms. Igaku-Shoin, New York, Tokyo
3. McBee K, Bickham JW (1988) Petrochemical related DNA damage in wild rodents detected by flow cytometry. Bull Environ Contam Toxicol 40:343–349
4. Bickhman JW, Hanks BG, Smolen MJ et al (1988) Flow cytometric analysis of low-level radiation exposure on natural populations of slider turtles (*Pseudomys scripta*). Arch Environ Contam Toxicol 17:837–841
5. Lowcock LA, Sharbel TF, Bonin J et al (1997) Flow cytometry assay for in vivo genotoxic effect of pesticides in green frogs (*Rana clamintas*). Aquat Toxicol 30:241–255
6. Lamb T, Bickham JW, Lyne TB et al (1995) The slider turtle as an environmental sentinel: multiple tissue assay using flow cytometric analysis. Ecotoxicology 4:5–13
7. Dallas CE, Lingenfelser SF, Lingenfelser JT et al (1998) Flow cytometric analysis of erythrocyte and leukocyte DNA in fish from Chernobyl-contaminated ponds in the Ukraine. Ecotoxicology 7:211–219
8. Debenest T, Gagne F, Burgeot T et al (2013) DNA integrity assessment in hemocytes of soft-shell clams (*Mya arenria*) in the Saguenay Fjord (Quebec, Canada). Environ Sci Pollut Res 20:621–629
9. Bihari N, Mičić M, Batel R et al (2003) Flow cytometric detection of DNA cell cycle alterations in hemocytes of mussel (*Mytilus galloprovincialis*) off the Adriatic coast, Croatia. Aquat Toxicol 64:121–129
10. Bihari N, Hamer B, Jakšić Ž et al (2002) Application of alkaline elution, fast micromethod and flow cytometry in detection of marine contamination. Cell Mol Biol 48(4):373–377
11. Mičić M, Bihari N, Mlinarič-Raščan I (2004) Influence of herbicide, 2,4-dichlorophenoxy acetic acid, on hemocyte DNA of *in vivo* treated mussel. J Exp Mar Biol Ecol 311: 157–169
12. Mičić M, Bihari N, Labura Ž et al (2001) Induction of apoptosis in the blue mussel *Mytilus galloprovincialis* by tri-*n*-butyltin chloride. Aquat Toxicol 55:61–73
13. Elston RA, Kent ML, Drum AS (1988) Progression, lethality and remission of haemic neoplasia in the bay mussel *Mytilus edulis*. Dis Aquat Organ 4:135–142
14. Aardema MJ, Albertini S, Arni P et al (1998) Aneuploidy: a report of an ECETOC task force. Mutat Res 410:3–79
15. Wersto RP, Liblit RL, Koss LG (1991) Flow cytometric DNA analysis in human solid tumours: a review of the interpretation of DNA histograms. Progr Pathol 22:1085–1098

Chapter 3

Ultrasound Imaging of DNA-Damage Effects in Live Cultured Cells and in Brain Tissue

Hadi Tadayyon, Mehrdad J. Gangeh, Roxana Vlad, Michael C. Kolios, and Gregory J. Czarnota

Abstract

High-frequency ultrasound (>20 MHz) spectroscopy can be used to detect noninvasively DNA damage in cell samples in vitro, and in live tissue both ex vivo and in vivo. This chapter focuses on the former two aspects. Experimental evidence suggests that morphological changes that occur in cells undergoing apoptosis result in changes in frequency-dependent ultrasound backscatter. With advances in research, ultrasound spectroscopy is advancing the boundaries of fast, label-free, noninvasive DNA damage detection technology with potential use in personalized medicine and early therapy response monitoring. Depending on the desired resolution, parametric ultrasound images can be computed and displayed within minutes to hours after ultrasound examination for cell death.

Key words Cell death, Ultrasound, Cancer response imaging, Quantitative ultrasound

1 Introduction

1.1 Quantitative Ultrasound Spectroscopic Detection of DNA Damage

This chapter describes a novel ultrasound-based method of cell death detection in live cultured cells and excised tissue. The method is sensitive to many forms of cell death—key among these is apoptotic cell death. Multicellular organisms regulate the number of cells through mitosis and apoptosis. Apoptosis is a form of programmed cell death (suicide) which naturally occurs when a cell is not needed. Cancers can arise from a disruption in the regulation of apoptosis, resulting in uncontrolled cell growth. Apoptosis is a specialized form of cell death that is genetically controlled with unique morphological features at a cellular level. The process is initiated by chromatin aggregation forming a pyknotic nucleus that breaks down into small fragments. Through analysis of the frequency-dependent ultrasound backscatter signal, one can detect characteristic morphologically linked changes associated with apoptosis and mitotic arrest, both of which ultimately lead to cell death. High-frequency quantitative ultrasound studies (ultrasound

Vladimir V. Didenko (ed.), *Fast Detection of DNA Damage: Methods and Protocols*, Methods in Molecular Biology, vol. 1644, DOI 10.1007/978-1-4939-7187-9_3,

operating at center frequencies above 20 MHz) have demonstrated large increases in the ultrasound backscatter intensity (6–12 dB) coincident with cell death, in both cell aggregates [1–3] and preclinical experiments [4–6]. Extensive research carried out over two decades has demonstrated in vitro, ex vivo, and in vivo (preclinical and clinical studies) that cell death can be detected within 24 h of anticancer treatment administration including photodynamic therapy, radiotherapy, chemotherapy, and anti-vascular therapy [5–8]. Due to the involvement of more preparation steps (i.e., centrifugation) for cell cultures, cell death detection is more time consuming when compared to live tissues ex vivo and in vivo. Nevertheless, depending on the desired resolution and computing hardware available, a parametric ultrasound image mapping the cell death extent in a tissue image can be obtained now within 5 min (based on a 20 MHz ultrasound center frequency, 0.1 mm image resolution, ~5 × 5 mm region of interest, and using an Intel Core i7 processor). Increasing the resolution to as high as 0.03 mm could result in an approximate quadruple increase in computation time. Whereas current ultrasound data collection and analysis technology has not been computationally optimized, considerable computational performance improvements can be achieved by utilizing graphics processing units of the computing hardware and parallelizing any matrix-based computations. This chapter will describe methods and protocols for detecting ultrasonically DNA damage in vitro in cell samples, and ex vivo in rat brains.

1.2 Background

1.2.1 Ultrasound Imaging

Sound is an indispensable mode of communication which is used in our daily lives. It is essentially a pressure wave produced by a mechanical vibration. In humans, the vibration is created by the vocal cords in the larynx and in stereo speakers, by an oscillating electromagnet. Ultrasound refers to sound with frequencies above the upper audible limit of human hearing, which is 20 kHz. An ultrasound image is generated by spatially mapping the amplitude of the backscattered (reflection) pressure from different tissue interfaces as sound propagates through the tissue. The strength of the reflected signal and its frequency content depend on the properties of these structures (e.g., size, density and compressibility). Thus, ultrasound is sensitive to the biomechanical properties of tissues which often change with disease and cell death processes.

Ultrasound is commonly used clinically in screening, diagnosis, and image-guided procedures, due to its relatively low cost, short imaging time, good image resolution, ionizing radiation-free nature, and intrinsic tumor contrast which does not require injection of any contrast agents. Brightness mode (B-mode) imaging, which provides grayscale intensity images generated by reflections from different tissues and their interfaces, is the standard mode of ultrasound imaging used by sonographers and radiologists for biomedical diagnosis and screening. Diagnostic ultrasound imaging is commonly performed using ultrasound frequencies in the range of 1–10 MHz, with applications in breast, prostate, liver, and echocardiography. These frequencies can ensure the visualization of deep body structures of interest with a reasonable resolution (approximately 0.15–1.5 mm). Within the range of frequencies of 10–60 MHz, high-frequency ultrasound offers a high spatial resolution up to 25 μm and can enable longitudinal studies in mice with applications in developmental biology and cardiovascular disease [9–11], tumor growth [12, 13], angiogenesis, and assessing anti-angiogenic drug effects [14]. A caveat to B-mode imaging is the fact that the frequency-dependent information from tissue echo signals is lost during conversion of the raw ultrasound backscatter radiofrequency (RF) signal to B-mode images. Ultrasound spectrum analysis is an ultrasound signal analysis technique applied to the raw RF ultrasound data to extract parameters that can characterize various tissue abnormalities such as those of the prostate, lymph nodes, eye, as well as the myocardium [15–18].

Experimentally, ultrasound imaging and spectrum analysis techniques have been applied by our group [1, 4, 5, 19] to detect cell death in cell samples and tissues exposed to cancer therapies. Specifically, within the frequency range of 10–60 MHz, ultrasound imaging and tissue characterization techniques have been used to detect cell death, in vitro with cell samples exposed to the chemotherapeutic drug cisplatinum [1] ex vivo [1, 20] and in vivo using tumor mouse models exposed to cancer therapies [5–7].

The resolution of ultrasound imaging, even at the frequencies of 10–60 MHz is not as great as the resolution offered by optical microscopy. However, ultrasound confers the advantage of a relatively low attenuation, permitting deeper penetration of ultrasound waves into the tissue (i.e., 1–5 cm deep compared to micron-range depths obtained using optical microscopy). Rather than resolving individual structures (e.g., cells and nuclei) the spectral parameters computed from the analysis of ultrasound signals provide information about the statistical properties of tissue structure at the scale of a cell. In the next section, a brief description of ultrasound spectrum analysis methods to characterize cellular death in cell samples and biological tissues is presented.

1.2.2 Quantitative Ultrasound Spectroscopic Detection of Cell Death Based on Scattering Properties

In quantitative ultrasound spectroscopy (QUS), frequency-domain analysis is performed on the raw ultrasound backscatter signal, referred to as the RF signal, in order to extract information about tissue microstructure otherwise not resolvable by conventional gray-scale (B-mode) images. The most basic QUS parameters that can be extracted from RF data are spectral slope (SS), spectral intercept (SI), and mid-band fit (MBF). These parameters have been used as a means to differentiate pathological tissue, mainly cancer, from normal tissue in sites such as the prostate, lymph node, eye, and myocardium [15–18] and later used to characterize the response of a cancerous tissue to cancer therapy [5, 6, 21].

In an initial study using high-frequency (40 MHz) ultrasound, where acute myeloid leukemia cells were treated with cisplatinum (a chemotherapeutic agent), a 25-to-50 fold increase in the US backscatter intensity was observed in apoptotic cells compared to viable cells. This observation led to the hypothesis that the cell nucleus is the source of ultrasound scattering, and that it is the morphological changes occurring in the nucleus during apoptosis that causes such ultrasound backscatter changes. This hypothesis was investigated by Taggart et al. [22], where they demonstrated experimentally that cells with larger nuclear diameters express higher integrated backscatter coefficients. Ultrasound signal changes coincident with cell death were later quantified by QUS parameters including MBF and SS, both of which demonstrated an increase as a result of cell death [21]. This observation was experimentally determined to be related to the nuclear condensation and fragmentation that occur during apoptosis. The same pattern of changes observed in vitro was observed in animal cancer models in vivo, resulting from photodynamic therapy [5], or radiation therapy [6] of mouse tumor xenografts. Furthermore, QUS parameters overlaid on standard ultrasound images provide a delineation of areas of positive response to cancer therapy, as previous studies have shown a good spatial correlation between high-intensity areas in QUS images of integrated backscatter (IBS; backscatter signal integrated over frequency, similar to MBF) and positive stain areas in terminal deoxynucleotidyl transferase dUTP nick end labeling (TUNEL) stained histology images [6].

More recently, QUS parameters such as MBF and SI have demonstrated utility in differentiating therapy responding breast cancer patients from nonresponding ones during neoadjuvant chemotherapy [23]. Alternatively, broader frequency bandwidths permit estimation of advanced QUS parameters such as average scatterer diameter (ASD) and average acoustic concentration (AAC), which are derived by fitting a scattering model to the RF data [24]. These parameters have been demonstrated to be effective in differentiating mouse carcinomas from rat fibroadenomas [25], low grade breast tumors from medium to high-grade ones [26], and chemotherapy responsive breast tumors from nonresponsive ones within 4 weeks of a several-month-long treatment [27].

2 Materials

2.1 Ultrasound Imaging System

A VisualSonics VS40B high-frequency ultrasound device (Visual Sonics Inc., Toronto, Ontario, Canada) was used to collect ultrasound images and radio-frequency data from cell samples and rat brain tissue ex vivo, as shown in Fig. 1a. We used such instruments configured with single element transducers and a "digital RF"

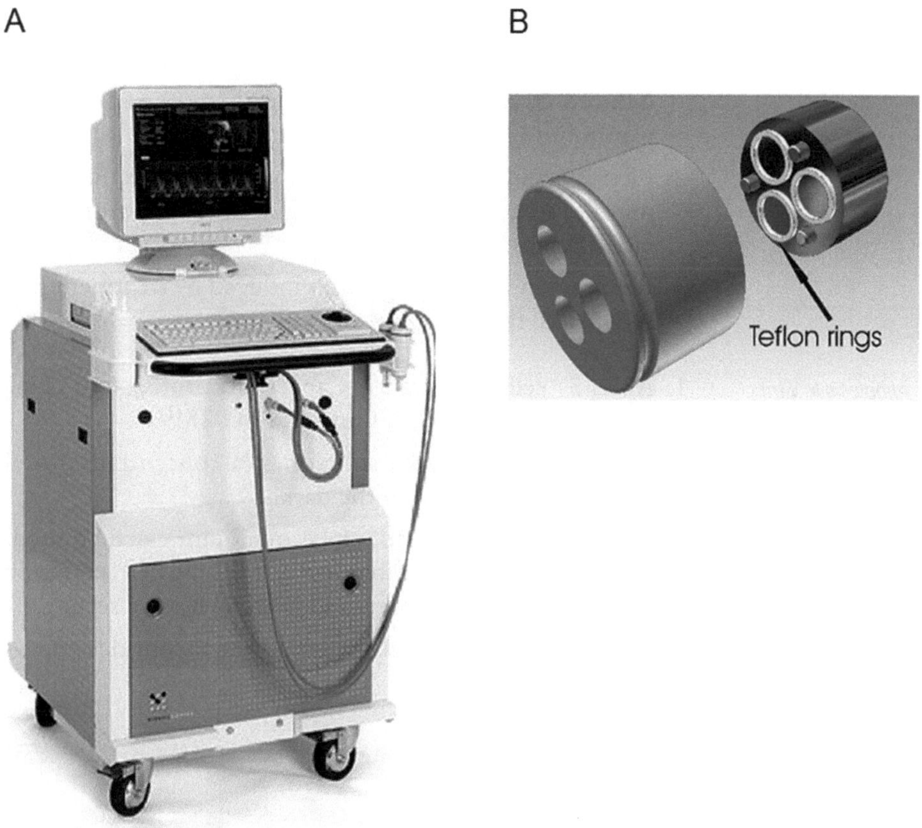

Fig. 1 (**a**) The Vevo 770 high-resolution micro-imaging system for real-time small animal imaging. Similarly to its VS40B predecessor, the VEVO 770 can be used to visualize and quantify small animal anatomical, hemodynamic and therapeutic targets, and intervention effects with resolution of up to ~30 μm. Using these instruments, the region of therapeutic intervention can be monitored in the same animal over time (i.e., before, during, and after therapy). The Vevo 770 has an automated 3D acquisition capability. (**b**) Photograph of the sample holder. The cell sample holder is made of stainless steel and has two parts: a flat bottom very finely polished to have the surface roughness much smaller than the ultrasound wavelength and a stainless-steel disk with three cylindrical holes cut through it, each 8 mm × 7 mm (diameter × height). The flat bottom is attached with screws to the stainless-steel disk and the bottoms of the wells are cushioned with Teflon rings. These Teflon rings ensure that the content of the wells is not spilled between compartments during the sample centrifugation. Two samples can thus be prepared once by simultaneous centrifugation of each in a sample well. The other well left empty serves as calibration reference. This figure is reprinted with permission from: DNA Damage Detection In Situ, Ex Vivo and In Vivo: Methods and Protocols edited by Vladimir V. Didenko (Methods in molecular biology; 682), 2011 [34]

module capable of acquiring, digitizing, and exporting raw radiofrequency data for spectral analysis. The transducer choice depends on system capabilities, application, and the specimen to be imaged. For most of the applications described here one can use two or three transducers to image a single specimen in order to cover a larger range of frequencies.

2.2 DNA Damage

The model systems described in this chapter are:

1. Acute myeloid leukemia cells (AML-5) in which apoptosis is induced by treating cells with the chemotherapeutic drug cisplatinum [1].
2. Hep-2 (epidermoid carcinoma of the larynx) cells in which mitotic arrest/catastrophe is induced by treating cells with radiotherapy [28].
3. Rat brain in which apoptosis is induced by treatment with photodynamic therapy [1].

2.2.1 Apoptosis In Vitro

1. AML-5 cells: (Ontario Cancer Institute, 610 University Avenue, Toronto, ON, Canada M5G 2M9).
2. Cisplatinum: the stock solution at a concentration of 1 mg/mL should be stored in the dark.
3. Growth media: 150 mL alpha minimum essential media supplemented with antibiotics (100 mg/L penicillin and 100 mg/L streptomycin) and 5% fetal bovine serum.
4. Flat-bottom cryo-tubes (1 mL) or a custom-designed sample cell holder.

2.2.2 Apoptosis Ex Vivo

1. Male Fischer rats.
2. Photofrin II (QLT, Canada).
3. Laser (632 nm) with an optical power irradiance of 100 mW/cm^2.
4. For anesthesia purposes: 100 mg/kg Ketamine, 5 mg/kg Xylazine, and 1 mg/kg Acepromazine in saline.

   These three drugs are under the regulations of the Controlled Drugs and Substances Act.

2.2.3 Other Modalities of Cell Death, Mitotic Arrest/Catastrophe

1. Hep-2 cell culture (ATCCmerican Type Culture Collection, Manassas, USA).
2. Growth medium: 150 mL alpha-minimum essential media supplemented with antibiotics (0.1% gentamicin) and 10% fetal bovine serum.
3. Trypsin (1× Trypsin–EDTA solution).

2.3 Flow Cytometry

1. Fluorescent activated cell sorting Caliber (BD Biosciences, Mississauga, Ontario, Canada) employing FL-2A and FL-2W channels.

2.4 Analysis of Ultrasound Backscatter Signals

We use a custom-made MATLAB program for ultrasound spectral analysis. The analysis requires two sets of data: sample RF data (raw ultrasound data capture of tissue) and reference RF data. The TEP previously described was used as reference RF data in both animal and human imaging studies. Details about ultrasound signal analysis are provided in Subheading 3.2.

3 Methods

DNA damage (leading to apoptosis, mitotic arrest, or mitotic catastrophe) was induced by exposing cell samples, rat brains, and tumor xenografts to different anticancer therapies. Users are free to use other materials and reagents to induce apoptosis and mitotic arrest/catastrophe. For example, in some of our previous experiments, AML cells were treated with colchicine that arrests cells in mitosis [4, 29] and Hep-2 cells were exposed to camptothecin that induces mainly apoptosis in this cell line [28].

For in vitro imaging of cell samples an adequate number of cells is required to provide a sample of packed cells of at least 150 mm^3 in volume. Potentially, any cell line may be used. To date, we have worked with leukemia, melanoma, breast, cervix, prostate and head and neck cancer cell cultures. Cell lines that do not grow in liquid suspension culture, once confluent, need to be trypsinized in order to free the cells from the growth surface of the flask. In the experiments described in this chapter, we have used AML and Hep-2 cell lines to demonstrate apoptosis and mitotic arrest/catastrophe. These protocols can be adapted for other cell culture systems. For any experimental time or condition, experiments are typically completed in triplicate.

3.1 Apoptosis In Vitro

1. AML cells obtained from frozen stock samples are cultured in a humidified atmosphere containing 5% CO_2 at 37 °C using the growth medium described in Subheading 2.2.1.
2. To induce apoptosis, AML-5 cells are exposed to cisplatinum dissolved in the growth medium at a concentration of 10 μg/mL for 0, 6, 12, 24, and 48 h. This drug is a DNA intercalater that causes a p53-dependent apoptosis in this cell line [30]. To confirm and quantify the presence of apoptosis we use different methods: examination of cell cultured using phase-contrast microscopy, examination of hematoxylin and eosin stained samples or cell cycle analysis using flow-cytometry.
3. Viable and treated cells are washed in phosphate-buffered saline and subsequently centrifuged in three steps at 800 × *g*, 1500 × *g*, and 1900 × *g*. After each centrifugation the cell culture media is aspirated and cells are washed in phosphate-buffered saline. The last centrifugation takes place in the cryo-tubes or in a custom sample holder (Fig. 1b) resulting in a small solid sample of packed cells that emulates a segment of tissue.
4. The sample holder or the cryo-tube is immersed in phosphate-buffered saline which acts as a coupling medium through which ultrasonic waves propagate. Each of the wells or tubes containing the viable and treated cells is imaged with high-frequency ultrasound.

   Representative results are presented in Figs. 3 and 4 and are discussed in **Notes 1–6**.

3.2 Other Forms of Cell Death

1. Hep-2 cells obtained from a frozen stack are cultured in a humidified atmosphere containing 5% CO_2 at 37 °C using the growth medium described in Subheading 2.2.3.
2. The flasks containing the cell samples are irradiated with 8 Gy radiation dose using a Faxitron Cabinet X-ray System.
3. Structural changes that are characteristic of apoptotic and mitotic response (i.e., increase in cellular and nuclear size, membrane ruffling, cytoplasmic vacuolization, nuclear fragmentation and condensation and formation of apoptotic and mitotic bodies) are used as an indication of response to radiotherapy. These structural changes are observed in the Hep-2 cell line typically 48 h after exposure to radiotherapy.
4. Viable and control cell samples are harvested by trypsinization and centrifuged at 800 × *g*. To prepare the cells for high-frequency imaging, cells are subsequently centrifuged and washed in phosphate-buffered saline as described at Subheading 3.1 from **steps 3** to **5**.

   Representative results are presented in Fig. 5 and discussed in **Notes** 7 and **8**.

3.3 Apoptosis Ex Vivo

1. Male Fischer rats are treated with 12.5 mg/kg of Photofrin II injected intraperitoneally and kept in the dark for 24 h prior to drug activation.
2. The animals are anesthetized using ketamine injected intraperitoneally using a dose of 2–4 mg/kg.
3. A 5.5 mm craniotomy is performed in each side of the rat's skull with a mechanical surgical drill, taking care to avoid significant mechanical stress to the underlying cortex.
4. The exposed brain is treated for 30 s using a red laser light with a wavelength of 632 nm and a spot size of 3 mm in diameter. This spot size allows simultaneous visualization of the treated region next to an untreated region in the 4 mm field of view of the high-frequency ultrasound scanner.
5. Several treatment irradiance powers of 1, 3, 5, and 17 J/cm^2 are used. In order to minimize cerebral swelling post-therapy, the irradiance power of 3 J/cm^2 was selected for further study. The optical irradiance power at the dural surface is 100 mW/cm^2.
6. To study the treatment effects at different time points the animals are sacrificed at 1.5, 3, and 24 h after the application of photodynamic therapy.

   Representative results are presented in Fig. 6 and are discussed in **Notes 9** and **10**.

3.4 Ultrasound Data Collection and Spectral Analysis

1. All cell samples and live tissues were imaged at room temperature immersed in phosphate-buffered saline (PBS).
2. For ultrasound data collection in vitro and ex vivo, we used Visual Sonics VS40B high-frequency ultrasound device equipped with a 20 MHz focused transducer (20 mm focal length, 8 mm aperture diameter, −6 dB bandwidth of 11–28 MHz) and a 40 MHz focused transducer (9 mm focal length, 6 mm aperture diameter, −6 dB bandwidth of 25–55 MHz) for in vitro and ex vivo imaging, respectively.
3. Radio-frequency data are typically collected from five different scan planes from cell samples and from five to ten different scan planes when using larger samples such as rat brain tissue. Each plane contains 15–40 8-bit radio-frequency lines sampled at 500 MHz and separated by a distance equal to the beamwidth of the transducer used in the respective application.
4. The region of interest (ROI) chosen to calculate the average ultrasound parameters is 4–6 mm wide and 1 mm in height centered at the transducer focus. These regions of interest are selected where the images appear to be homogeneous with no interfaces or large echoes.
5. Each ROI is multiplied by a Hamming window in order to suppress spectral side lobes, or in other words, suppress the nonrelevant frequency components. Subsequently, a Fourier transform is computed in order to obtain the power spectrum. Next, the system-dependency is removed by normalizing the sample power spectrum to the power spectrum of a flat quartz RF signal under the same imaging conditions. Linear regression analysis is then applied to the normalized backscatter power spectra in order to compute the spectral parameters as presented in Fig. 2b. The ultrasound integrated backscatter is similar to the mid-band fit described by the spectrum analysis framework developed by Lizzi et al. [31, 32] and is determined by the effective scatterer size, concentration, and difference in acoustic impedance between the scatterers and surrounding medium. The spectral slope can be related to the effective scatterer size [32] (i.e., an increase in the spectral slope corresponds to a decrease in the average scatterer size) and spectral intercept depends on effective scatterer size, concentration, and relative acoustic impedance (Fig. 2). Further details on the theoretical and signal analysis considerations and how spectral parameters are related to tissue microstructure can be found elsewhere [31, 32].
6. A discussion on ultrasound imaging of cell death and future developments is provided in **Notes 11** and **12**.

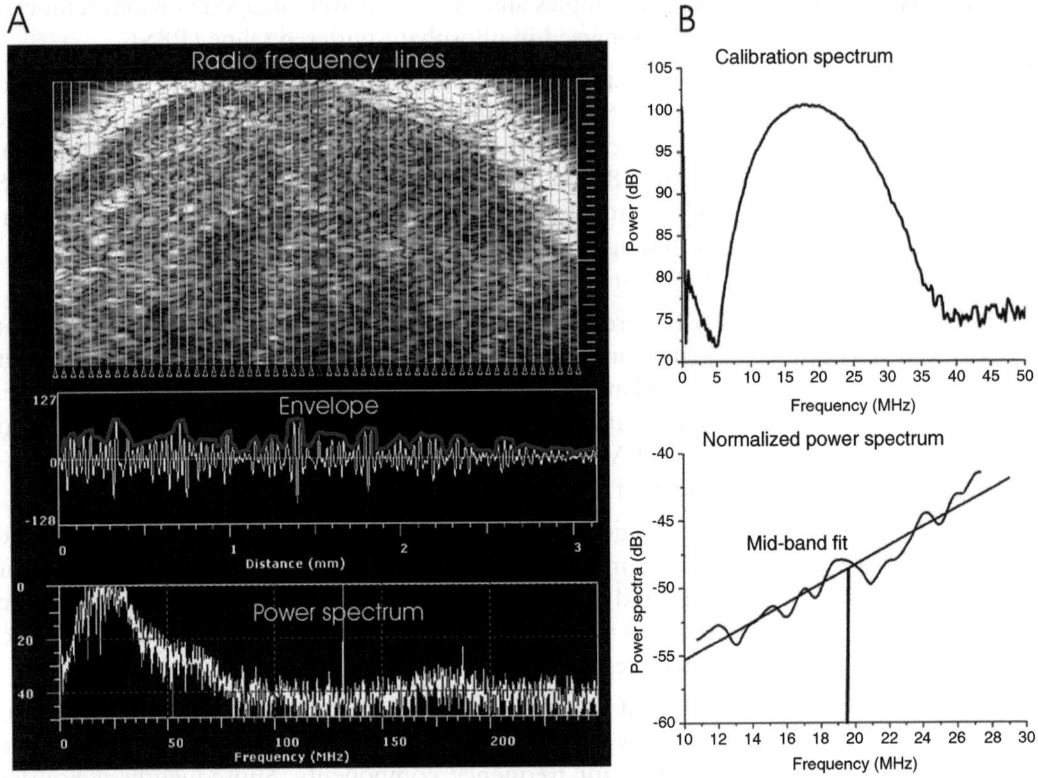

Fig. 2 (**a**) This panel describes the acquisition of radiofrequency data. The *upper panel* represents a selection from an ultrasound image with the scanning radio-frequency lines (pre-image data) detected by the ultrasound transducer; the *middle panel* represents the radio frequency signal with the outline of the signal (envelope) corresponding to one of the scan lines. In order to create the ultrasound images typically displayed by the conventional ultrasound machines, these raw radio-frequency data undergo different signal processing steps, e.g., envelope detection. The *bottom panel* represents the power spectrum corresponding to the same scan line from the *upper panel*. (**b**) This panel describes the method of ultrasound spectrum analysis. The *upper panel* represents the calibration spectrum that is a power spectrum measured from a flat quartz immersed in water and placed at the transducer focus. The normalized power spectrum (*lower panel*) is calculated by dividing the power spectrum calculated from a region of interest by the calibration spectrum. Linear regression analysis is then applied to the resulting spectrum. The spectral slope (dB/MHz) and the spectral intercept (dB) are the slope and the intercept of this line, respectively. The ultrasound-integrated backscatter (dB) is calculated by integrating the normalized power spectrum over the transducer's −6 dB bandwidth. This is typically similar to the mid-band fit that is the value of the regression line calculated at a frequency corresponding to the middle of the bandwidth. This figure is reprinted with permission from: DNA Damage Detection In Situ, Ex Vivo and In Vivo: Methods and Protocols edited by Vladimir V. Didenko (Methods in molecular biology; 682), 2011 [34]

3.5 Flow-Cytometry, Cytology, and Histology Analysis

3.5.1 Flow-Cytometry Cell-Cycle Analysis

1. Cells are harvested, washed twice in fluorescent activated cell sorting buffer (phosphate-buffered saline/0.5% bovine serum albumin), resuspended in 1 mL of FACS buffer and fixed for 1 h on ice in 3 mL of ice-cold 70% ethanol. Cells are then washed once before resuspending them in 500 μL of fluorescent activated cell sorting buffer supplemented with 40 μg/mL RNAse A and 50 μg/mL propidium iodine. Cells are then incubated at room temperature for 30 min in the dark before being analyzed using a flow cytometer (described in Subheading 2.3).
2. The results are represented as percent number of cells found in different phases of the cell cycle. This classification is dependent on DNA content and helps to identify the forms of cell death following each of the applied therapies (Figs. 4 and 5c).

3.5.2 Cytology and Histology Analysis

1. Cell samples, rat brains, and tumor xenografts are fixed in 10% neutral-buffered formalin and embedded in paraffin.
2. A TUNEL assay is used to histologically detect apoptosis in the rat brain treated with photodynamic therapy. In this type of stain, the free DNA ends resulted from chromatin fragmentation in apoptotic nuclei are labeled with a green fluorescent stain and cytoplasm is marked red with a propidium iodide counterstain (Fig. 6)

4 Notes

Ultrasound images of apoptotic samples, corresponding histology and measurements of DNA content are presented in Figs. 3 and 4.

1. In order to minimize the work required for cell culture and preparation of the experiments, cells that grow in liquid suspension and have a short doubling time are preferred to adherent cells. For example, four to six flasks of adherent cells (175 cm^2—the area of one flask with 75% cell confluence) are needed to provide a suitable volume of cellular material for a cell sample. We have preferred AML cells for our proof-of-principle experiments, because they grow in suspension, have a relative short doubling time of 12 h, and do not need trypsinization. Furthermore, they are a good experimental model for apoptosis studies because the cells have a round regular shape with a large nucleus (~8.5 μm vs. cellular size ~10 μm). This makes it relatively easy to follow nuclear morphological changes under light microscopy.
2. AML cells kept a long time in cell culture (>4 weeks) may change their phenotype due to genetic instability and increase in size. This can affect their response to cytotoxic agents (e.g., cisplatin) and affect their ultrasound spectra.

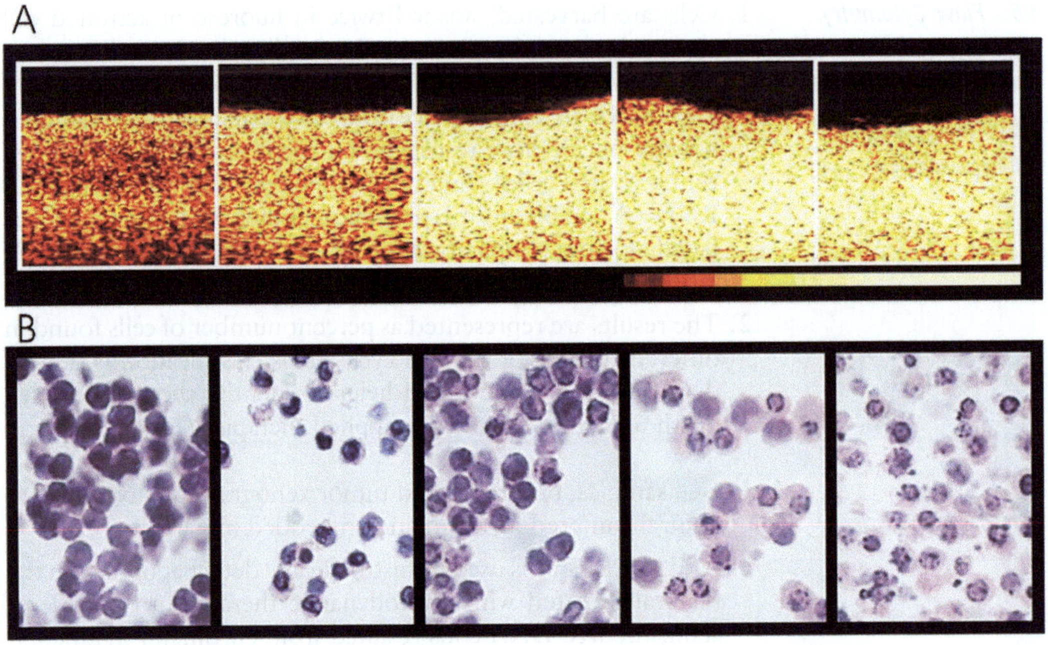

Fig. 3 Ultrasound imaging of apoptosis, in vitro, and corresponding histology. *Top panels* from *left* to *right* represent ultrasound images corresponding to cell samples treated with cisplatinum for 0, 6, 12, 24, and 48 h. The ultrasound images indicate a progressive increase in ultrasound backscatter, observed as an increase in the brightness of the cell samples exposed to chemotherapy. The speckle pattern (the *bright* and *dark dots*) does not correspond to individual cells. This speckle results from the constructive and destructive interference of multiple small waves reflected by tissue scattering structures. The color-bar at the *bottom right* of the figure corresponds to pixel values ranging from 0 (black) to 256 (white). *Bottom panels* show representative images of hematoxylin and eosin-stained histology corresponding to each ultrasound image. The treated samples demonstrate the typical characteristics of apoptosis, nuclear condensation, and fragmentation that become more prominent with longer time exposure to the drug. The field of view in each hematoxylin and eosin image is approximately 50 μm. This figure is reprinted with permission from: In situ detection of DNA damage: methods and protocols edited by Vladimir V. Didenko (Methods in molecular biology; 203), 2002 [35]

3. Cells need to be carefully washed with phosphate-buffered saline and cell culture medium has to be removed before ultrasound imaging. Cell culture media has tensioactive properties (tends to form bubbles) that can result in significant artifacts when imaging with ultrasound.
4. The volume of a cell sample has to be reduced from 600 mL (four flasks of AML cells) to ~0.45–1.00 mL (the volume of a well in the custom sample holder or cryo-tube) during the three steps of cell centrifugation. At this stage the cell material is very viscous and can be difficult to pipette. The last centrifugation should result in a 0.10–0.20 mL solid cell sample. The recommended height of a sample is at least ~3 mm to facilitate artifact-free ultrasound data collection. However, the height of the sample depends on the transducer specifications and needs to be optimized accordingly.

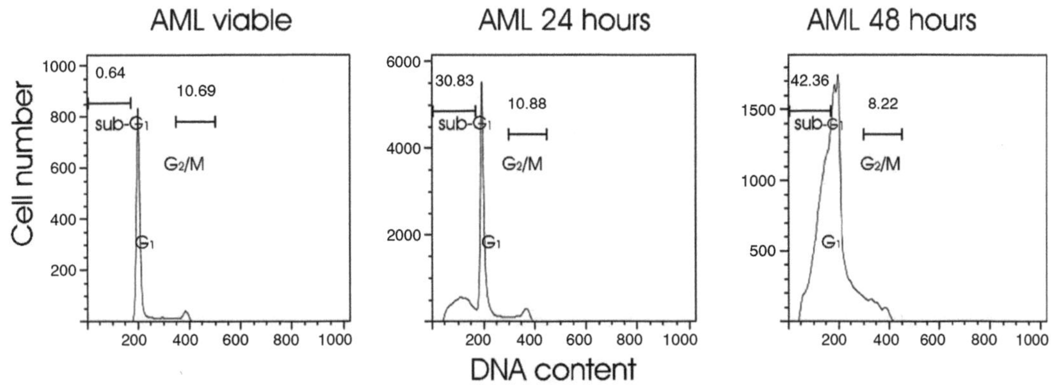

Fig. 4 Cell cycle analysis corresponding to AML cells treated with cisplatin for 0, 24, and 48 h. The sub-G1 fractions are identified as apoptotic cells because the nucleus becomes fragmented during apoptosis. The sub-G1 peak can represent, in addition to apoptotic cells, mechanically damaged cells and cell fragments resulting from advanced stages of cell death. The sub-G1 peak increased from 0.6% in viable AML cell samples to 30.8 and 42.4% after 24 and 48 h, respectively, exposure to cisplatin. The G1 peak represents phenotypic normal cells and the G2/M peak is identified as cells in mitosis and mitotic arrest. No significant changes are observed in the G2/M peak with the treatment (10.7% at 0 h; 10.9% at 24 h, and 8.2% at 48 h). This figure is reprinted with permission from: DNA Damage Detection In Situ, Ex Vivo and In Vivo: Methods and Protocols edited by Vladimir V. Didenko (Methods in molecular biology; 682), 2011 [34]

5. To avoid trapping small bubbles at the surface of the cell sample the well or cryo-tube containing the cell sample has to be gently overfilled with phosphate-buffered saline before placing it into a larger volume of the same solution for ultrasound imaging.
6. For imaging cell samples in vitro, we have designed a custom sample holder (Fig. 1b). This optimizes the quantity and geometry of cells available for an imaging session and allows data collection in a certain manner for the calculation of other ultrasounds parameters (e.g., speed of sound and attenuation coefficient). For rigorous calculations of cell pellet properties, this (or a similar) setup is required for the appropriate corrections. An alternative technique is to prepare the samples in flat-bottom cryo-tubes. However, these may be too tall (height of the tube > transducer focal length) to access cell samples with certain transducers and may have to be height adjusted.

Ultrasound images of mitotic arrest/catastrophe, corresponding histology and cell cycle analysis are presented in Fig. 5.

7. To induce other forms of cell death we have used head and neck cancer lines that die predominantly by mitotic arrest/catastrophe after exposure to radiotherapy [3]. This modality of cell death is characterized by enlarged cells and nuclei, in contrast to nuclear condensation and fragmentation and

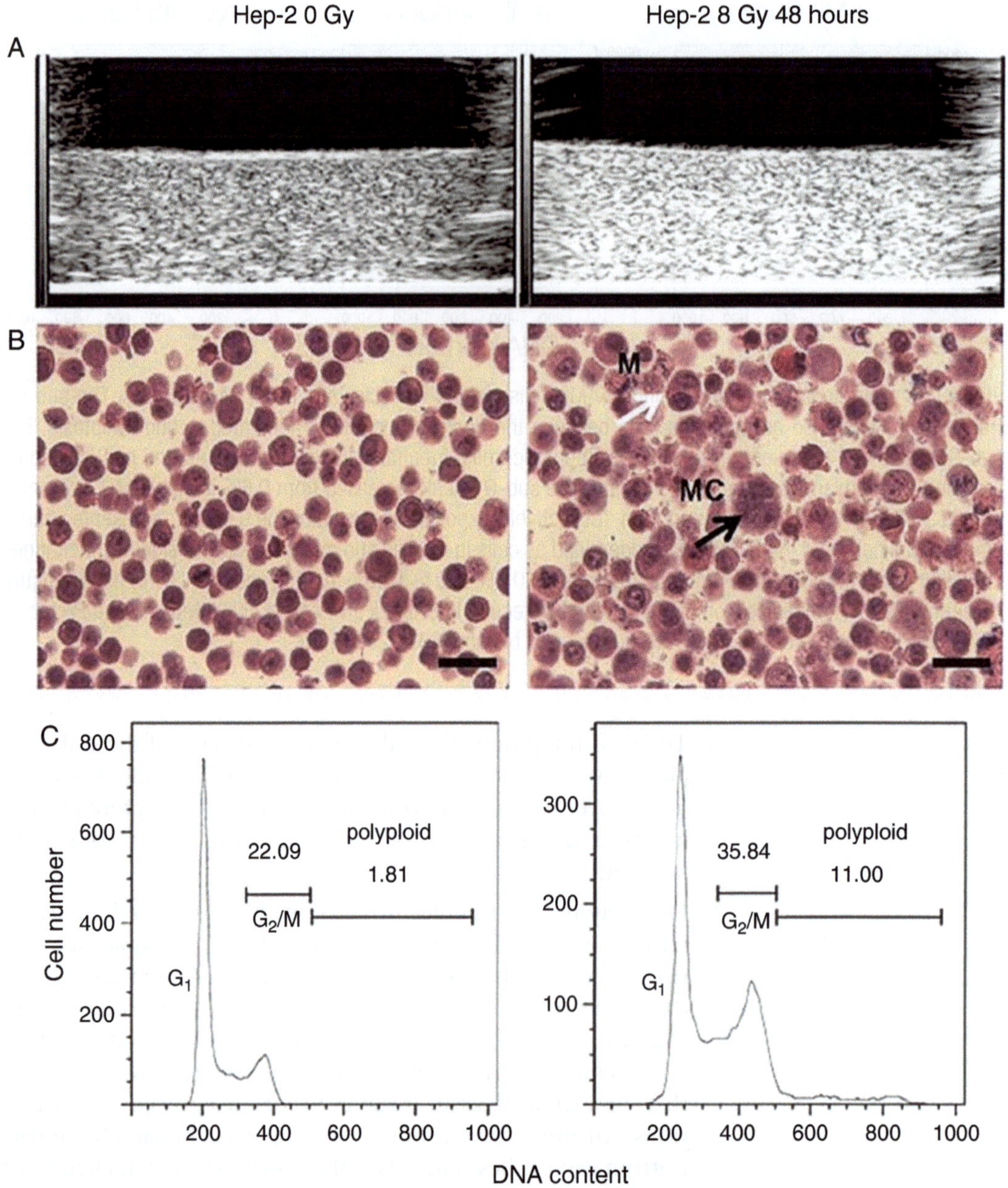

Fig. 5 Ultrasound imaging of mitotic arrest/catastrophe, in vitro, corresponding histology and cell cycle analysis. (**a**) The ultrasound images indicate an increase in ultrasound backscatter of the cell sample treated with radiotherapy. (**b**) The corresponding hematoxylin and eosin image demonstrate cells and nuclei with larger size consistent with cell death by mitotic arrest (*white arrow*) and mitotic catastrophe (*black arrow*). The scale bar represents 40 μm. (**c**) The corresponding cell cycle analysis indicates an increase in the mitotic cell fraction quantified by the G_2/M fraction (22.1% at 0 Gy and 35.8% at 8 Gy) and a sixfold increase in the polyploid cell fraction (1.8% at 0 Gy and 11.9% at 8 Gy) after radiotherapy. No sub-G_1 fractions are detectable. This figure is reprinted with permission from: DNA Damage Detection In Situ, Ex Vivo and In Vivo: Methods and Protocols edited by Vladimir V. Didenko, 2011 [34]

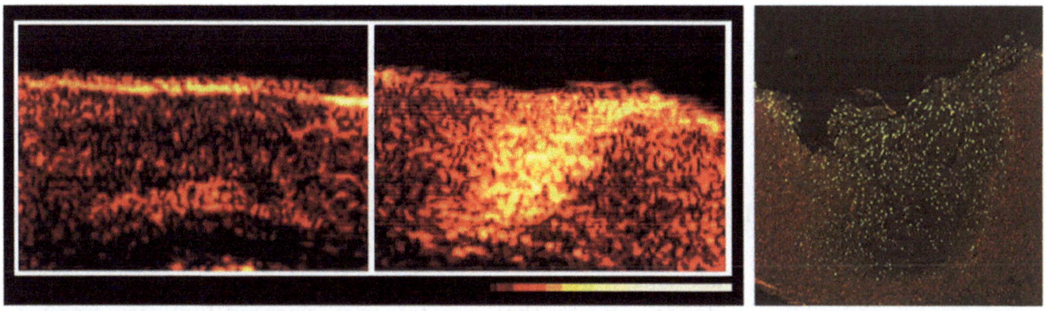

Fig. 6 Ultrasound imaging of apoptosis, ex vivo, in response to photodynamic therapy. The *left* and *central panels* are the ultrasound images corresponding to the control and treated tissue, respectively. The wedge-shaped *yellow area* in the *central panel* presents an increase in the ultrasound backscatter corresponding to the treated region. The color-bar at the *bottom right* of the figure corresponds to pixel values ranging from 0 (black) to 256 (white). The *right panel* presents fluorescence microscopy assay for apoptosis corresponding to the ultrasound image from the *central panel*. The intact cells stain red with propidium iodide, whereas the DNA-ends of fragmented DNA in apoptotic cells stain green with fluorescein dye. The *wedge-shaped region stained green* in the histology corresponds to the wedge-shaped hyperechoic area in the ultrasound image of the treated brain tissue. This figure is reprinted with permission from: In situ detection of DNA damage: methods and protocols edited by Vladimir V. Didenko (Methods in molecular biology; 203), 2002 [35]

cellular shrinking characteristic of apoptotic cell death. Minor evidence of cell undergoing apoptosis is observed in the Hep-2 cell culture examined under light microscopy, but no significant apoptotic fraction is measured by cytometry (Fig. 5).

8. The Hep-2 cell line has a cell cycle of up to 48 h and up to 50% of the cells express mitotic arrest/catastrophe around this time after treatment. Light microscopy images of cell culture do not reveal significant amount of damage at earlier time points. Based on these observations, we consider that 48 h is a good time to observe a significant effect in ultrasound backscatter since keeping the treated cells longer than 2 days in the cell culture would allow the surviving cells to further divide decreasing the chance to effectively image early cell death.

9. Figure 6 illustrates the use of high-frequency ultrasound imaging to detect apoptosis that occurred ex vivo in a rat model treated with photodynamic therapy. Similar results have been obtained in vivo on skin rat treated with photodynamic therapy [1].

10. Any type of treatment able to induce apoptosis in tissue may result in changes in ultrasound images and ultrasound parameters in the treated tissue compared with the same region before treatment. However, it is recommended that the tissue chosen for investigation to have a relative homogeneous structure because ultrasound backscatter is also sensitive to the degree of randomness in scatterer arrangement. In preliminary

experiments, we have measured no increase in ultrasound backscatter from cell samples with large degree of randomness in the positions of nuclei, although flow-cytometric measurements indicated that a large majority of cells (>50%) underwent death in these cell samples [33].

11. Although some of the changes in spectral parameters estimates can be interpreted in terms of changes in nuclear size and acoustic impedance, recent theoretical and experimental evidence indicates that increases in nuclear randomization may, in addition, influence significantly the magnitude of ultrasound backscatter [33]. In these circumstances an understanding of the relative contribution of each of these nuclear changes (acoustic impendence versus randomization) to ultrasound scattering is essential in order to accurately quantify cell death. Ongoing studies in our laboratory aim to precisely and reliably measure acoustic properties as a function of treatment [21].
12. In conclusion, ultrasound imaging and quantitative ultrasound methods can be used as a method of detecting, noninvasively, cell structural changes resulting from DNA damage in a variety of settings. In this chapter, we demonstrated the detection of such DNA damage in vitro and ex vivo following cell death after exposure to various anticancer therapies. High-frequency ultrasound imaging and quantitative ultrasound methods are able to provide very specific information about changes in cellular structure (i.e., differentiate between apoptotic and mitotic cell death). The next chapter will describe the extension of quantitative ultrasound detection of cell death to the lower-frequency range which has the potential to noninvasively detect cell death in numerous clinically relevant scenarios including detecting tumor responses to specific therapies and guiding tissue biopsies and thus, help in developing tools for the clinical management of cancer treatment. Since monitoring therapy response is becoming an essential component of drug development, the ultrasound methods described here can be used in preclinical studies to assess responses to new-experimental anticancer therapies in animal cancer models. Imaging of therapy responses in animal cancer models permits repeated assessments of the tumor and may provide spatial and temporal information regarding target organs and the heterogeneity of the response. This information can be correlated with tumor biopsies and histopathological methods to further provide correlative therapy response endpoints related to ultrasound methods.

Acknowledgments

M.J.G. holds the Natural Sciences and Engineering Research Council of Canada Post-doctoral Fellowship. G.J.C. holds a University of Toronto James and Mary Davie Chair in Breast Cancer Imaging and Ablation. Funding for these projects was provided by the Terry Fox Foundation.

References

1. Czarnota GJ, Kolios MC, Abraham J et al (1999) Ultrasound imaging of apoptosis: high-resolution noninvasive monitoring of programmed cell death in vitro, in situ and in vivo. Br J Cancer 819(3):520–527
2. Tunis AS, Czarnota GJ, Giles A, Sherar MD, Hunt JW, Kolios MC (2005) Monitoring structural changes in cells with high-frequency ultrasound signal statistics. Ultrasound Med Biol 31:1041–1049
3. Vlad RM, Alajez NM, Giles A, Kolios MC, Czarnota GJ (2008) Quantitative ultrasound characterization of cancer radiotherapy effects in vitro. Int J Radiat Oncol Biol Phys 72:1236–1243
4. Kolios MC, Czarnota GJ, Lee M, Hunt JW, Sherar MD (2002) Ultrasonic spectral parameter characterization of apoptosis. Ultrasound Med Biol 28:589–597
5. Banihashemi B, Vlad R, Debeljevic B, Giles A, Kolios MC, Czarnota GJ (2008) Ultrasound imaging of apoptosis in tumor response: novel preclinical monitoring of photodynamic therapy effects. Cancer Res 68:8590–8596
6. Vlad RM, Brand S, Giles A et al (2009) Quantitative ultrasound characterization of responses to radiotherapy in cancer mouse models. Clin Cancer Res 15(6):2067–2074
7. Sadeghi-Naini A, Falou O, Tadayyon H, Al-Mahrouki A, Tran W, Papanicolau N, Kolios MC, Czarnota GJ (2013) Conventional frequency ultrasonic biomarkers of cancer treatment response in vivo. Transl Oncol 6:234–243
8. Kim HC, Al-Mahrouki A, Gorjizadeh A, Sadeghi-Naini A, Karshafian R, Czarnota GJ (2014) Quantitative ultrasound characterization of tumor cell death: ultrasound-stimulated microbubbles for radiation enhancement. PLoS One 9:18–20
9. Foster FS, Pavlin CJ, Harasiewicz KA, Christopher DA, Turnbull DH (2000) Advances in ultrasound biomicroscopy. Ultrasound Med Biol 26:1–27
10. Foster FS, Zhang MY, Zhou YQ et al (2002) A new ultrasound instrument for in vivo microimaging of mice. Ultrasound Med Biol 28:1165–1172
11. Foster FS, Zhang M, Duckett AS, Cucevic V, Pavlin CJ (2003) In vivo imaging of embryonic development in the mouse eye by ultrasound biomicroscopy. Invest Ophthalmol Vis Sci 44:2361–2366
12. Graham KC, Wirtzfeld LA, MacKenzie LT, Postenka CO, Groom AC, MacDonald IC, Fenster A, Lacefield JC, Chambers AF (2005) Three-dimensional high-frequency ultrasound imaging for longitudinal evaluation of liver metastases in preclinical models. Cancer Res 65:5231–5237
13. Cheung AMY, Brown AS, Hastie LA, Cucevic V, Roy M, Lacefield JC, Fenster A, Foster FS (2005) Three-dimensional ultrasound biomicroscopy for xenograft growth analysis. Ultrasound Med Biol 31:865–870
14. Goertz DE, Yu JL, Kerbel RS, Burns PN, Foster FS (2002) High-frequency Doppler ultrasound monitors the effects of antivascular therapy on tumor blood flow. Cancer Res 62:6371–6375
15. Coleman DJ, Lizzi FL, Silverman RH, Helson L, Torpey JH, Rondeau MJ (1985) A model for acoustic characterization of intraocular tumors. Invest Ophthalmol Vis Sci 26: 545–550
16. Feleppa EJ, Kalisz A, Sokil-Melgar JB et al (1996) Typing of prostate tissue by ultrasonic spectrum analysis. IEEE Trans Ultrason Ferroelectr Freq Control 43:609–619
17. Yang M, Krueger TM, Miller JG, Holland MR (2007) Characterization of anisotropic myocardial backscatter using spectral slope, intercept and midband fit parameters. Ultrason Imaging 29:122–134

18. Mamou J, Coron A, Oelze M et al (2011) Three-dimensional high-frequency backscatter and envelope quantification of cancerous human lymph nodes. Ultrasound Med Biol 37(3):345–357
19. Czarnota GJ, Kolios MC, Vaziri H, Benchimol S, Ottensmeyer FP, Sherar MD, Hunt JW (1997) Ultrasonic biomicroscopy of viable, dead and apoptotic cells. Ultrasound Med Biol 23:961–965
20. Vlad RM, Czarnota GJ, Giles A, Sherar MD, Hunt JW, Kolios MC (2005) High-frequency ultrasound for monitoring changes in liver tissue during preservation. Phys Med Biol 50:197–213
21. Brand S, Weiss EC, Lemor RM, Kolios MC (2008) High frequency ultrasound tissue characterization and acoustic microscopy of intracellular changes. Ultrasound Med Biol 34: 1396–1407
22. Taggart LR, Baddour RE, Giles A, Czarnota GJ, Kolios MC (2007) Ultrasonic characterization of whole cells and isolated nuclei. Ultrasound Med Biol 33:389–401
23. Sadeghi-Naini A, Papanicolau N, Falou O et al (2013) Quantitative ultrasound evaluation of tumor cell death response in locally advanced breast cancer patients receiving chemotherapy. Clin Cancer Res 19:2163–2174
24. Insana MF, Wagner RF, Brown DG, Hall TJ (1990) Describing small-scale structure in random media using pulse-echo ultrasound. J Acoust Soc Am 87:179–192
25. Oelze ML, O'Brien WD Jr, Blue JP, Zachary JF (2004) Differentiation and characterization of rat mammary fibroadenomas and 4T1 mouse carcinomas using quantitative ultrasound imaging. IEEE Trans Med Imaging 23(6):764–771
26. Tadayyon H, Sadeghi-Naini A, Czarnota GJ (2014) Noninvasive characterization of locally advanced breast cancer using textural analysis of quantitative ultrasound parametric images. Transl Oncol 7:759–767
27. Sannachi L, Tadayyon H, Sadeghi-Naini A, Tran W, Gandhi S, Wright F, Oelze M, Czarnota G (2014) Non-invasive evaluation of breast cancer response to chemotherapy using quantitative ultrasonic backscatter parameters. Med Image Anal 20:224–236
28. Brand S, Solanki B, Foster DB, Czarnota GJ, Kolios MC (2009) Monitoring of cell death in epithelial cells using high frequency ultrasound spectroscopy. Ultrasound Med Biol 35: 482–493
29. Czarnota GJ, Kolios MC, Hunt JW, Sherar MD (2002) Ultrasound imaging of apoptosis. DNA-damage effects visualized. Methods Mol Biol 203:257–277
30. Zamble DB, Lippard SJ (1995) Cisplatin and DNA repair in cancer chemotherapy. Trends Biochem Sci 20:435–439
31. Lizzi FL, Greenebaum M, Feleppa EJ, Elbaum M, Coleman DJ (1983) Theoretical framework for spectrum analysis in ultrasonic tissue characterization. J Acoust Soc Am 73:1366–1373
32. Lizzi FL, Astor M, Liu T, Deng C, Coleman DJ, Silverman RH (1997) Ultrasonic spectrum analysis for tissue assays and therapy evaluation. Int J Imaging Syst Technol 8:3–10
33. Vlad R, Orlova V, Hunt J, Kolios M, Czarnota G (2008) Changes measured in the backscatter ultrasound signals during cell death can be potentially explained by an increase in cell size variance. Ultrason Imaging 29:256
34. Didenko VV (2011) DNA damage detection in situ, ex vivo, and in vivo: methods and protocols, Methods in Molecular Biology. Springer, New York
35. Didenko VV (2002) In situ detection of DNA damage: methods and protocols, Methods in Molecular Biology. Springer, New York

Chapter 4

Ultrasound Imaging of Apoptosis: Spectroscopic Detection of DNA-Damage Effects In Vivo

Hadi Tadayyon, Mehrdad J. Gangeh, Roxana Vlad, Michael C. Kolios, and Gregory J. Czarnota

Abstract

In this chapter, we describe two new methodologies: (1) application of high-frequency ultrasound spectroscopy for in vivo detection of cancer cell death in small animal models, and (2) extension of ultrasound spectroscopy to the lower frequency range (i.e., 1–10 MHz range) for the detection of cell death in vivo in preclinical and clinical settings. Experiments using tumor xenografts in mice and cancer treatments based on chemotherapy are described. Finally, we describe how one can detect cancer response to treatment in patients noninvasively early (within 1 week of treatment initiation) using low-frequency ultrasound spectroscopic imaging and advanced machine learning techniques. Color-coded images of ultrasound spectroscopic parameters, or parametric images, permit the delineation of areas of dead cells versus viable cells using high ultrasound frequencies, and the delineation of areas of therapy response in patient tumors using clinically relevant ultrasound frequencies. Depending on the desired resolution, parametric ultrasound images can be computed and displayed within minutes to hours after ultrasound examination for cell death. A noninvasive and express method of cancer response detection using ultrasound spectroscopy provides a framework for personalized medicine with regards to the treatment planning of refractory patients resulting in substantial improvements in patient survival.

Key words Quantitative ultrasound, Breast cancer response prediction, Texture analysis, Machine learning, Computer-aided theragnosis

1 Introduction

1.1 Cancer Therapy Response Assessment and the Role of Ultrasound

Traditionally, cancer response to therapy is assessed based on a reduction in the sum of the largest diameters of target lesions or the largest diameter of unifocal disease [1]. However, clinically detectable reduction in tumor size does not typically occur until several weeks to months into treatment. Consequently, imaging assessments of tumor biology and biochemistry have led to the discovery of novel biomarkers that can provide earlier indications (within days) of tumor response to therapy [2]. Research in early detection of breast cancer response to anticancer therapy has led to

Vladimir V. Didenko (ed.), *Fast Detection of DNA Damage: Methods and Protocols*, Methods in Molecular Biology, vol. 1644, DOI 10.1007/978-1-4939-7187-9_4, © Springer Science+Business Media LLC 2017

discoveries in both image-based and chemical-based biomarkers. In a prospective clinical study, Chang et al. monitored levels of apoptotic index (AI) and Ki-67 in breast cancer patients undergoing chemotherapy through flow cytometric and immunocytochemical analyses of fine needle aspiration samples obtained from the breast tumors [3]. Whereas AI represents the number fraction of apoptotic cells identified through terminal deoxynucleotidyl transferase deoxyuridine-triphosphatase nick end labeling (TUNEL), staining for DNA fragmentation, Ki-67 represents cancer cell proliferation. In that study, an increase in AI after 1–3 days, and a decrease in Ki-67 after 21 days, all significant, were found in responding patients compared to nonresponding ones. In another study by Nishimura et al. [4], a higher Ki-67 index was found to be associated with poorer disease-free survival of breast cancer patients.

As for image-based markers, diffusion-weighted MRI (DW-MRI) has been demonstrated clinically to predict response of breast tumors as early as after one cycle of chemotherapy. It is used to measure the apparent diffusion coefficient corresponding to changes in the Brownian motion of water in the tumor tissue, which is thought to increase in responding tumors due to a decrease in tumor cellularity [5]. Tumors are known to have higher glucose metabolism than other tissues. On this basis, in nuclear imaging, a longitudinal positron emission tomography (PET) imaging study on breast cancer used fluorodeoxyglucose (FDG) contrast agent to enhance the tumor region and track its metabolism during chemotherapy treatment [6]. The study demonstrated that responding tumors could be detectable after one cycle of chemotherapy. Additionally, diffuse optical imaging (DOI) studies on breast cancer have demonstrated a significant increase in hemoglobin concentration, water percentage, and tissue optical index, in responding patients as early as 1 week after chemotherapy treatment initiation [7]. The utility of these modalities in the clinic, however, is limited due to long wait times and the requirement of contrast agents in the case of MRI, poor resolution and limited penetration in the case of DOI, and health concerns over the repeated use of radioactive material in the case of PET.

As described in the previous chapter, ultrasound has multiple advantages such as being relatively low cost, having short imaging time, high resolution, radiation-free nature, and intrinsic tumor contrast which does not require injection of any contrast agents. For background on quantitative ultrasound (QUS) imaging of cell death, please refer to the previous chapter, Subheading 1.2.

In the previous chapter, we described classical QUS parameters that, depending on the ultrasound frequency of interrogation, can be correlated to cell death, including spectral slope (SS), spectral intercept (SI), mid-band fit (MBF), and integrated backscatter (IBS). Additionally, advanced QUS parameters based on scattering models such as average scatterer diameter (ASD) and average acoustic concentration (AAC) have been used to characterize mouse models of breast cancer, malignant versus non-malignant lymph nodes, and more recently to monitor the response of breast cancer patients to chemotherapy [8–10].

Whereas in the previous chapter high-frequency (above 20 MHz) ultrasound and QUS analysis were demonstrated to be effective in detecting cell death in vitro and ex vivo, recent studies by our group have shown that such techniques can also be extended to a lower—clinically relevant—frequency (6 MHz), permitting in vivo detection of cell death in preclinical and clinical settings [10–17]. In ultrasound imaging, a lower center frequency corresponds to higher penetration depth at the cost of lower image resolution. At 6 MHz, the axial resolution is approximately 250 μm and penetrations as deep as 7 cm can be achieved, making transducers with this frequency suitable for in vivo imaging in patients.

1.2 Computer-Aided Theragnosis

Recent advances in QUS methods have motivated the design of computer-aided theragnosis (CAT) systems to reliably classify cancer patients as responders or nonresponders early after the start of treatment. We have developed a CAT system recently [16] using QUS methods in conjunction with state-of-the-art texture analysis and advanced machine learning techniques that can predict the response of LABC patients to neoadjuvant chemotherapy (NAC) in the first week of their several-month treatment with an accuracy of 85%. The CAT system consists of several major components including feature extraction using texture methods, measuring the dissimilarities between the "pre-" and "mid-treatment" scans using a kernel-based metric, learning from imbalanced data, and a supervised learning algorithm to classify the patients to responders/nonresponders.

2 Materials

2.1 Ultrasound Imaging Systems for Small Animals

For high-frequency data acquisition, a Vevo 770 small animal imaging system (Visual Sonics Inc., Toronto, Ontario, Canada) was used with a single-element transducer, RMV 710B. This system has seen widespread adoption by many small animal imaging facilities. We use such instruments configured with single-element transducers and a "digital RF" module capable of acquiring, digitizing, and exporting raw radio-frequency data for spectral analysis. New high-frequency linear array transducers (frequencies from 9 to 70 MHz) optimized for specific research applications have been recently released commercially (Vevo 2100). Such instruments are termed "micro-ultrasound" imaging systems and are dedicated to in vivo longitudinal imaging studies for small animal phenotyping. The transducer choice on this range of instruments depends on system capabilities, application, and the specimen to be imaged. For most of the applications described here, one can use two or three transducers to image a single specimen in order to cover a larger range of frequencies. For low-frequency data acquisition, a Sonix RP system (Ultrasonix, Vancouver, British Columbia, Canada) was used with a linear array transducer (Fig. 1a and b).

2.2 Ultrasound Imaging System for Patients

Patient imaging can be performed using a clinical ultrasound imaging system capable of RF data acquisition and storage. For breast imaging, we used a Sonix RP equipped with a 6 MHz center frequency linear array transducer (L14-5/60 W). Prior to scanning, an ultrasound gel was applied to the area of the breast to be scanned. A picture of the system is presented in Fig. 1a and c.

2.3 Tissue Equivalent Phantom

A tissue equivalent phantom (TEP) used as a calibration step was constructed in-house, using agar medium and spherical glass beads adapted from [18]. The glass beads were purchased from Cospheric and had a size range of 53–63 μm with density of 2.52 g/cc (Cospheric SLGMS-2.52 27–30 μm, Santa Barbara, CA). The bead radius was optically measured to be 20 ± 3 μm. Glass beads were suspended with a concentration of 2.2 g/L of molten agar medium (2% agar dissolved in distilled water), and the phantom was left to rotate in a rotisserie overnight to allow the beads to mix uniformly in the gel during the solidification process.

2.4 DNA Damage

The models systems described in this chapter are:

1. Human tumor xenografts grown in mice, in which apoptosis is induced by treatment with human-mimicking chemotherapy.
2. Human breast cancer tumors (in patients).

2.4.1 Chemotherapy in Mice

1. Severe combined immunodeficient mice.
2. Human breast cancer cells (MDA-MB-231; American Type Culture Collection (ATCC), Manassas, VA).
3. RPMI 1640 cell culture medium (Thermo Fisher Scientific, Waltham, MA USA) for growing the tumor supplemented with 10% fetal bovine serum and antibiotics (100 mg/L penicillin and 100 mg/L streptomycin).
4. For anesthesia purposes: 100 mg/kg Ketamine, 5 mg/kg Xylazine, and 1 mg/kg Acepromazine in saline.
5. Paclitaxel and Doxorubicin with biologically equivalent doses of 100 mg/m^2 and 50 mg/m^2 per mouse, respectively.

2.4.2 Chemotherapy in Patients

Patients follow standard clinical protocols for NAC and our imaging methods do not alter the clinical workflow. Chemotherapy is often administered using a combination of taxanes, anthracyclines, fluorouracil, and cyclophosphamide. Chemotherapy administration is typically fractionated (into cycles) in order to help patients recover from drug effects. The administration schedule is typically once every 2 or 3 weeks and the duration varies from 18 to 24 weeks.

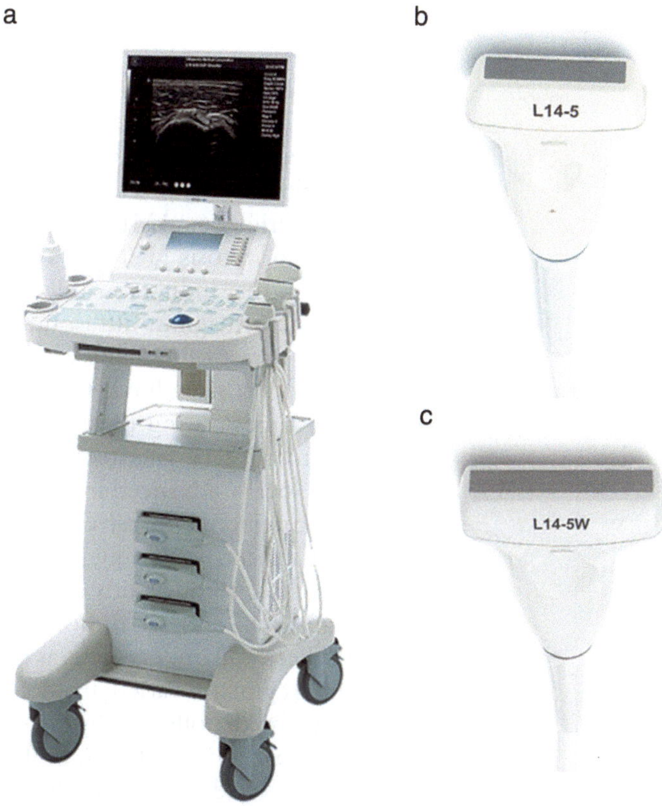

Fig. 1 A clinical ultrasound imaging system (Sonix RP system, Aanalogic Inc., Vancouver, Canada). The system is equipped with clinical (screening) and research modes and operates with array transducers (**a**). A small transducer (L14-5/38 mm) is used for mouse imaging, as shown in (**b**) and a large one (L14-5/60 mm) is used for patient breast imaging as shown in (**c**). Both transducers have 128 elements, whereas L14-5/38 is 38 mm wide and L14-5/60 is 60 mm wide. For animal imaging, scanning is mechanically actuated traversing over the tumor volume, whereas for patient imaging, scanning is performed in a free-hand manner, traversing over the affected area of the breast. This figure is reprinted with permission from: (**a**): http://www.ultrasonix.com/wikisonix/index.php/File:Sonixrp.jpg. (**b**): http://www.bkultrasound.com/ultrasonix/transducers/l14-538-linear. (**c**): http://www.bkultrasound.com/ultrasonix/transducers/l14-5w60-linear

2.5 Analysis of Ultrasound Backscatter Signals

We use a custom-made MATLAB program (The Mathworks Inc., Natick, MA, USA) for ultrasound spectral analysis. The analysis requires two sets of data: sample RF data (raw ultrasound data capture of tissue) and reference RF data. The TEP previously described was used as reference RF data in both animal and human imaging studies. Details about ultrasound signal analysis are provided in Subheading 3.2.

3 Methods

DNA damage (leading to apoptosis, mitotic arrest, or mitotic catastrophe) was induced by exposing mice to chemotherapy. Users are free to use other materials and reagents to induce apoptosis and mitotic arrest/catastrophe. For example, in some of our previous experiments, AML cells were treated with colchicine that arrests cells in mitosis [19, 20] and Hep-2 cells were exposed to camptothecin that induces mainly apoptosis in this cell line [21].

For clinical imaging, ultrasound RF data were collected from the affected breast of patients with locally advanced breast cancer (LABC) prior to neoadjuvant chemotherapy treatment initiation and at four times during the course of treatment—weeks 1, 4, 8, and prior to surgery (mastectomy/lumpectomy). Patients diagnosed with locally advanced invasive breast cancer, including invasive ductal carcinoma, invasive lobular carcinoma, and other forms of invasive cancer, including all grades, were referred from the diagnostic clinic to our study. This included patients with tumors larger than 5 cm and/or tumors with locoregional lymph node, skin, and chest wall involvement as per guidelines reported in [22]. Treatment regimens varied from 5-fluorouracil, epirubicin, and cyclophosphamide followed by docetaxol (FEC-D), to Adriamycin followed by paclitaxel (AC-T), or Taxol followed by Trastuzumab varying from weekly to tri-weekly cycles.

3.1 Apoptosis

1. In a humidified atmosphere containing 5% CO_2 at 37 °C and using the cell culture medium described previously, MDA-MB231 cells are cultured to a size of ~1.0×10^6 cells, which is injected intradermally into the left hind leg of each mouse. Primary tumors are measured weekly with a caliper and are allowed to develop for approximately 4 weeks until they reach a diameter of 6–10 mm.
2. A SCID mouse is anesthetized before imaging and hair over the tumor and surroundings is removed using a depilation cream.
3. Ultrasound RF data are collected via a mechanical sweep over the tumor-bearing leg (to construct a 3D ultrasound image) before treatment and a specified amount of time after treatment (in our experiments we had 4 treatment groups: 4, 12, 24, and 48 h posttreatment). Two sets of ultrasound data are acquired per tumor, high-frequency and low-frequency ultrasound data, using two systems—Vevo 770 and Ronix RP.
4. Prior to imaging, the tumor-bearing leg is placed in a water bath (*see* **Note 1**) which serves as the ultrasound coupling medium. A photograph of the setup is presented in Fig. 2.
5. Mice are treated with paclitaxel and doxorubicin through intravenous tail vein injection.

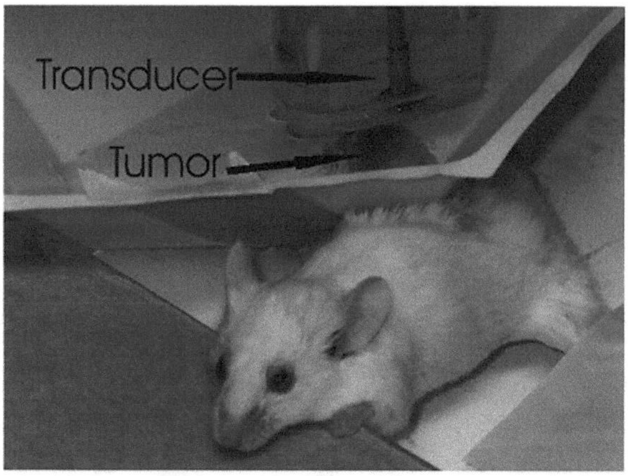

Fig. 2 A picture of the mouse placement for ultrasound imaging, showing the tumor-bearing leg pulled through a plastic container with the leg hole sealed by ultrasound gel and the container filled with distilled degassed water. The transducer is placed on the top of the leg with ~5 mm distance from the leg. This figure is reprinted with permission from: DNA Damage Detection In Situ, Ex Vivo and In Vivo: Methods and Protocols edited by Vladimir V. Didenko (Methods in molecular biology; 682), 2011 [23]

3.2 Ultrasound Data Collection and Spectral Analysis

In preclinical studies, the time between imaging sessions is often shorter (0, 4, 12, 24, and 48 h) compared to clinical studies (0, 1, 4, and 8 weeks). This is due to the fact that a one-time injection of a large chemotherapy dose is given to animals whereas for patients chemotherapy is given on a weekly or biweekly basis over a several-month period following clinical guidelines. However, in mice, we are interested in the cell death versus time profile and the length of time it takes to achieve maximum tumor cell death.

For ultrasound data collection of animal tumors, we used two systems—a high-frequency system and a low-frequency system. Measurements from the high-frequency system served as the gold standard, which were used to validate measurements made with the low-frequency system. The high-frequency system was a Vevo 770 system employing a 25 MHz transducer (RMV710B). RF data are typically collected from 10 to 20 different scan planes. Each plane consists of 250 RF lines sampled at 420 MHz and separated by a distance equal to the beam-width of the transducer used in the respective application. The transducer is positioned such that the tumor midline is at the depth of focus, which is 9 mm. For low-frequency ultrasound imaging of animals, we used the Sonix RP system and a 7 MHz center frequency linear array transducer (L14-5/38 mm) focused at 1.5 cm depth, with data sampled at 40 MHz, each plane consisting of 128 RF lines. Volumetric data were collected with scan plane separations of ~0.5 mm. For ultrasound data collected from patients with breast cancer, the Sonix RP system was used. The system was equipped with a 6 MHz linear array

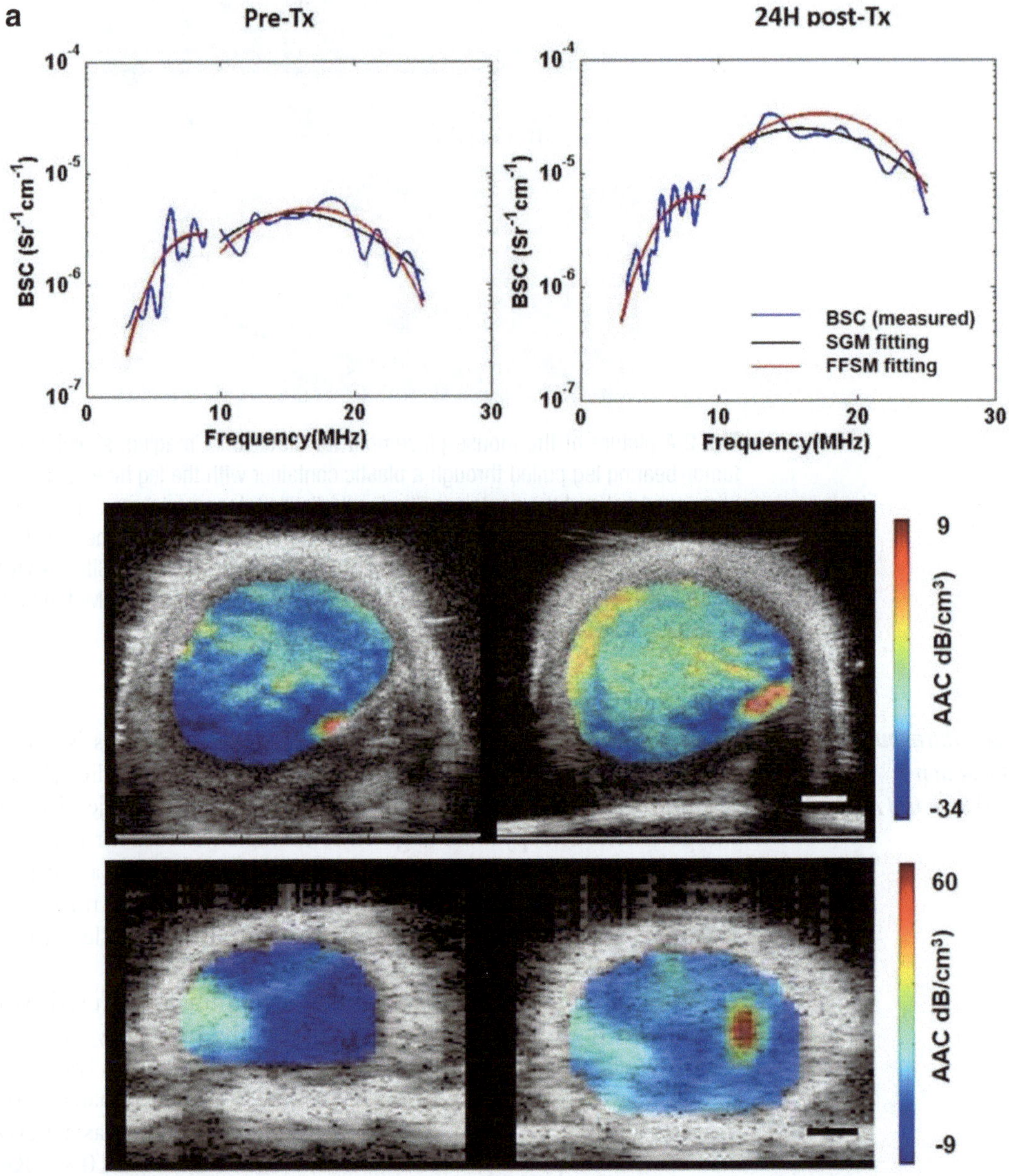

Fig. 3 (**a**) Top row: plots of BSC versus frequency from tumor of a mouse in vivo before (*left*) and 24 h (*right*) after chemotherapy drug injection. Measurements for 3–9 MHz were obtained from the L14-5 transducer with 6 MHz center frequency and measurements for 10–25 MHz were obtained from the RMV 710B transducer with 20 MHz center frequency. *Center row*: QUS images of AAC derived from high-frequency and low-frequency ultrasound data before and 24 h after chemotherapy treatment of mice. *Bottom row*: corresponding posttreatment TUNEL image shows large extent of cell death as indicated by the *brown* stains. Scale bars represent 2 mm. Adapted from [13]. (**b**) *Top row*: Plots of BSC versus frequency from the tumor of a LABC good response patient (*left*) and a poor response patient (*right*) prior to and 4 weeks after NAC initiation. *Bottom row*: Corresponding AAC images prior to and 4 weeks after NAC initiation. Scale bar: 1 cm. Adapted from [10]. This figure is reprinted with permission from: Quantification of Ultrasonic Scattering Properties of In Vivo Tumor Cell Death in Mouse Models of Breast Cancer by Hadi Tadayyon et al., 2015 [13] and Non-Invasive Evaluation of Breast Cancer Response to Chemotherapy using Quantitative Ultrasonic Bacscatter Parameters by Lakshmanan Sannachi et al., 2014 [10]

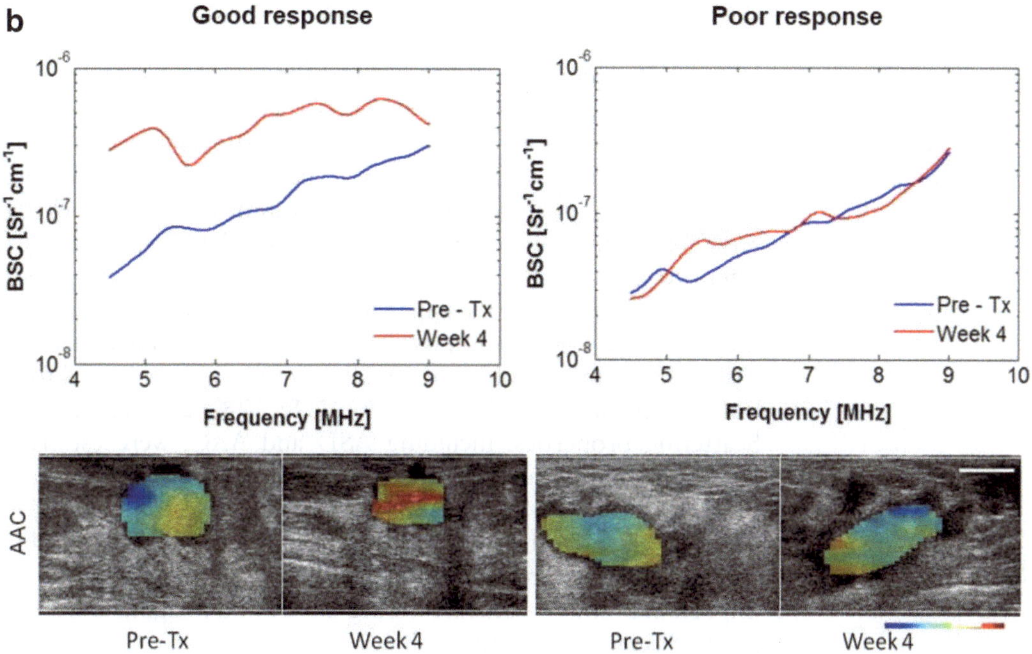

Fig. 3 (continued)

transducer (L14-5/60 W), with image planes consisting of 510 RF lines, sampled at 40 MHz. The ultrasound focus was placed at the midline of the tumor, which can be electronically adjusted from 10 to 60 mm.

Using the B-mode image for guidance, an ROI containing the tumor is selected for analysis. A B-mode image can be formed by taking the log-compressed absolute value of the Hilbert transform of the RF data [24]. The ROI should be a visually homogeneous containing the tumor core (not including skin or muscle surrounding it), as shown in Fig. 3. The ROI area is typically 2–4 × 2–4 mm^2 in mouse tumors and 2–4 × 2–4 cm^2 in human breast tumors (locally advanced breast cancers).

The ROI is segmented into sub-ROIs each 10 wavelengths long (~0.6 mm for high-frequency data for animals, and ~2 mm for low-frequency data for animals and for humans) in y direction (vertical direction; direction along ultrasound propagation) and 10 RF lines in *x* direction. The RF lines in each sub-ROI are multiplied by a Hanning window to suppress spectral lobes and each sub-ROI overlaps the adjacent sub-ROI by 80%.

From each sub-ROI, within the −6 dB frequency bandwidth of the system (the frequency range corresponding to 50% of the ultrasound spectral energy density), the average power spectrum is computed by taking the squared magnitude of the fast Fourier transform of each RF line and averaging the result across the RF lines. The average sample power spectrum is divided by the power spectrum of the TEP (2.3) in order to remove system dependent factors. This calibrated power spectrum is then multiplied by the backscatter coefficient (BSC) of the TEP in order to estimate the BSC of the sample. This is a well-established technique for estimating the BSC of a sample and is referred to as the reference phantom technique for BSC estimation [25]. The reference BSC was computed using the analytical model for the estimation of the backscattered pressure from solid spheres (glass beads) [26]. The newly obtained sample BSC was then compensated for frequency-dependent attenuation using the point-by-point compensation method [27]. For this purpose, the attenuation coefficients of the MDA-231 tumors and human LABC tumors were assumed to be 0.6 dB/cm-MHz and 1 dB/cm-MHz [28, 29], respectively. Scattering properties, including ASD and AAC, were estimated using the minimum of average squared deviation (MASD) fitting of the theoretical BSC and the measured BSC according to [27]. An illustration of the ROIs selected from tumors, corresponding to BSCs and QUS parametric images, is presented in Fig. 3. Some important points to consider during ultrasound imaging and analysis are discussed in **Notes 2–7**.

3.3 Computer-Aided Theragnosis System

Once a series of QUS parametric images have been obtained from a tumor ROI at specific times during treatment (whether it is from an animal or a human), they can be submitted to a computer-aided theragnosis (CAT) system that will identify tumor response to treatment. A schematic of the components of a CAT system is presented in Fig. 4. A CAT system consists of three main components:

3.3.1 Feature Extraction

The responses developed in tumors as a result of cancer treatment are often heterogeneous [30, 31]. This highlights potential advantages for textural analysis techniques, which can characterize these responses for a more accurate evaluation of therapeutic cancer response. The effectiveness of state-of-the-art texture methods such as local binary patterns (LBPs) [32], which are *predefined* binary operators, and texton-based methods [33, 34], which are based on data-driven codebooks learned from data has recently been demonstrated in cancer response-monitoring applications [16, 31, 35].

LBPs were first introduced by Ojala et al. [32] as a unified statistical and structural descriptor for texture analysis. The LBP method uses a histogram of predefined binary operators to estimate the underlying distribution of microstructural patterns such as edges, lines, corners, and spots in an image. The technique is based on the definition of local binary patterns that are circularly symmetric operators with certain radius R, which defines the spatial resolution of the operator, and with S equally spaced pixels on this circle, which determines the quantization of angular space. These operators are defined in such a way that the operator is invariant to local grayscale shifts.

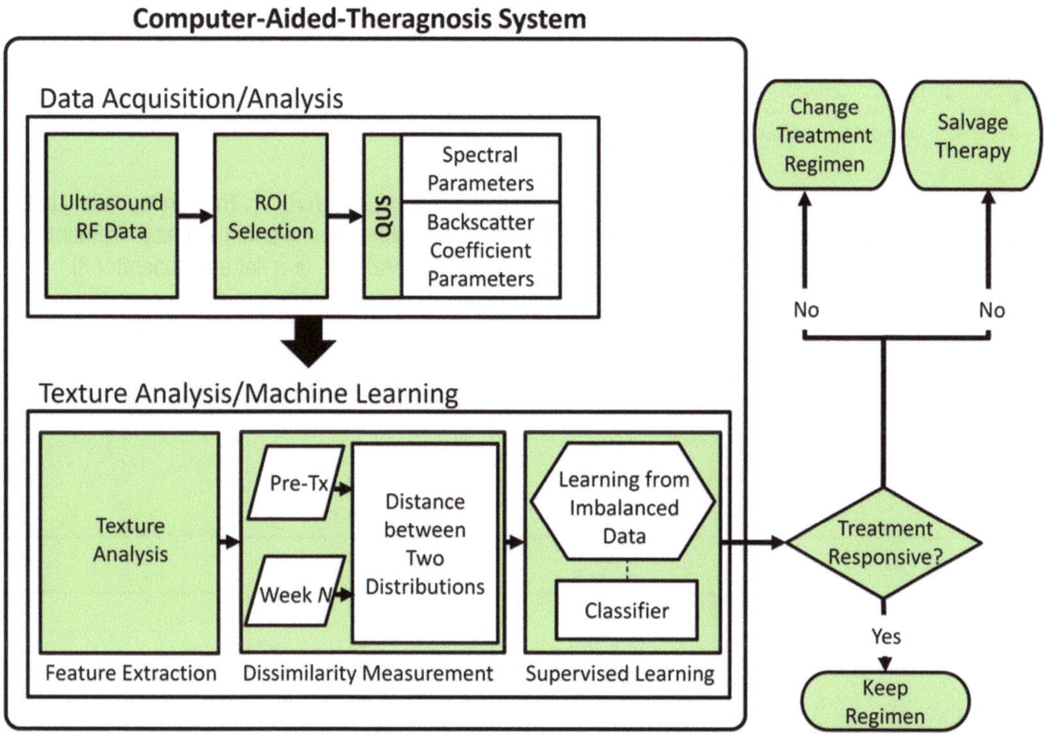

Fig. 4 The schematic of a computer-aided-theragnosis (CAT) system for the classification of patients as responders or nonresponders

Ojala et al. [32] introduced "uniform" LBPs and showed that they could be considered the fundamental property of image textures in the sense that the computation of LBPs responses on images mostly yielded "uniform" patterns (in some images more than 90%). Uniform LBPs have very few spatial transitions (less than three) and can detect microstructures such as edges, corners, spots, and flat areas (as shown in Fig. 5). The frequency of "uniform" patterns occurrence in an image has been demonstrated as a powerful texture feature descriptor [32]. Hence, the LBP approach is considered a combination of structural (by defining microstructural operators) and statistical (by using the histogram of these operators in the image as a feature descriptor) texture analysis technique.

Multiresolution analysis can be achieved by using the LBPs at multiple spatial radii. The most common LBP operators in the literature are: $LBP_{8,1}^{riu2}$, $LBP_{16,2}^{riu2}$, and $LBP_{24,3}^{riu2}$. The superscript in these operators stands for rotation invariant uniform 2 (i.e., uniform operator with maximum two spatial transitions), and the first index in the subscript represents S (the number of surrounding pixels), while the second one is R (the radius of the circular operator).

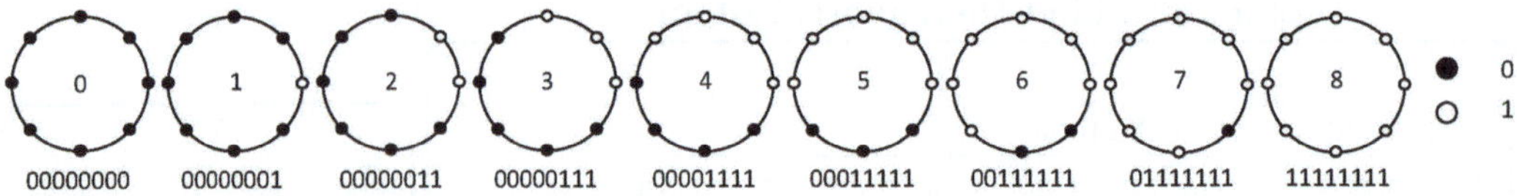

Fig. 5 The nine unique rotation invariant uniform 2 LBPs that can occur in $LBP_{8,1}^{riu2}$. The corresponding binary number is shown under each operator. Each binary operator corresponds to one primitive image element such as an edge (operators 2–6), corners (operators 1 and 7), a spot (operator 0), or a flat area (operator 8)

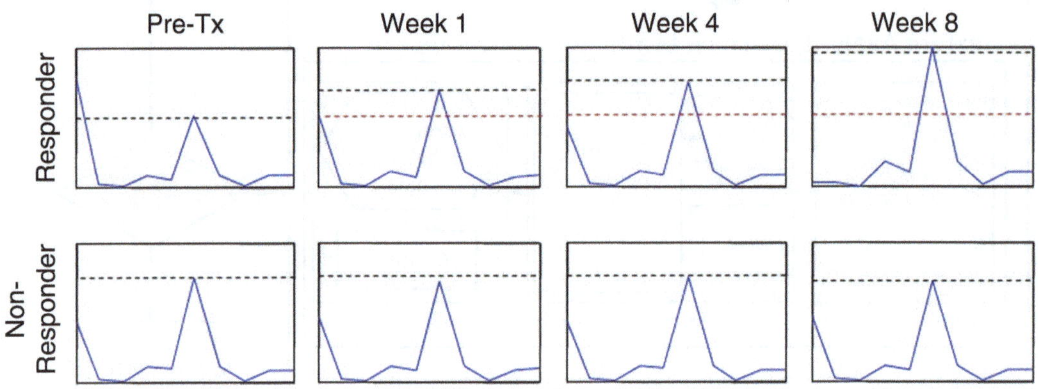

Fig. 6 The histograms of the LBPs ($LBP_{8,1}^{riu2}$) computed for a representative responder and nonresponder on MBF parametric maps over the course of treatment. As evident, the peak of the histograms of binary patterns increases for the responder over the course of treatment, indicating a change in the textural properties of the tumor image, whereas it remains almost unchanged for the nonresponder. This figure is reprinted with permission from Computer Aided Theragnosis Using Quantitative Ultrasound Spectroscopy and Maximum Mean Discrepancy in Locally Advanced Breast Cancer by MJ Gangeh et al., 2016 [16]

The application of the LBPs as textural feature descriptors in cancer response monitoring was first demonstrated in [16]. For instance, Fig. 6 illustrates the use of LBP during neoadjuvant chemotherapy to monitor patients with LABC tumors. Presented are the LBP histograms over the treatment period for a representative responding and nonresponding patient. The bins represent the LBP operators (depicted in Fig. 5). One can observe variations in the LBP histogram for the responder case, whereas there are little or no changes in that of the nonresponder case.

3.3.2 Dissimilarity Measurement

One main component of a CAT system is an efficient dissimilarity measure to correctly identify the distance of "pre-" and "mid-treatment" scans as an indication of treatment effectiveness. The main assumption here is that different modalities of cell death induced due to cancer therapies cause many morphological changes in tumor cells that will affect the bio-acoustic properties of the tumor [36, 37]. These changes are reflected in QUS parametric maps and can be quantitatively measured by deploying an appropriate dissimilarity measure. One proposal is to use a kernel-based metric such as maximum mean discrepancy (MMD) [38]. The MMD is based on the concept of Hilbert-Schmidt independence criterion (HSIC) [39] and was first introduced by Gangeh et al. to quantize cell death responses to cancer treatment in xenograft tumor models [40, 41]. This method is a distance measure, appropriate when there are multiple data samples available from two populations to be compared. It is a kernel-based measure, implying that its computation is reliant on inner products taken in a reproducing kernel Hilbert space (RKHS). By using a kernel function to nonlinearly transform input vectors into a different, possibly higher-dimensional feature space, and computing the population means in this new space, enhanced group separability is ideally obtained.

HSIC is a kernel-based measure to estimate the dependency of two random variables $\mathcal{X}$ and $\mathcal{Y}$. It computes the Hilbert-Schmidt norm of the cross-covariance operators in RKHS. For practical purposes, HSIC has to be estimated using a finite number of data samples. Considering $\mathcal{Z} = [(\mathbf{x}_1, \mathbf{y}_1), \ldots, (\mathbf{x}_n, \mathbf{y}_n)] \subseteq \mathcal{X} \times \mathcal{Y}$ as n independent observations drawn from $P_{\mathcal{X}\times\mathcal{Y}}$, an empirical estimate of HSIC can be computed as follows

$$\mathrm{HSIC}(\mathcal{Z}) = \frac{1}{(n-1)^2} tr(\mathrm{KHLH}), \tag{1}$$

where tr is the trace operator, H, K, $L \in \mathrm{R}^{n \times n}$, $K_{i,j} = k(\mathbf{x}_i, \mathbf{x}_j)$, $L_{i,j} = l(\mathbf{y}_i, \mathbf{y}_j)$, and $H = I - n^{-1}ee^{\mathrm{T}}$ (I is the identity matrix, e is a vector of n ones and hence, H is the centering matrix). It was shown in [38] that the MMD metric is closely related to the HSIC and can be computed to measure the distance between two populations x and y using

$$d_{\mathrm{HSIC}}(\mathbf{x},\mathbf{y}) = \left[\frac{1}{n^2} \sum_{i,j} K(\mathbf{x}_i,\mathbf{x}_j) - \frac{2}{nm} \sum_{i,j} K(\mathbf{x}_i,\mathbf{y}_j) + \frac{1}{m^2} \sum_{i,j} K(\mathbf{y}_i,\mathbf{y}_j) \right]^{\frac{1}{2}}, \tag{2}$$

where $\mathbf{x}_i$, $\mathbf{x}_j$, $\mathbf{y}_i$, and $\mathbf{y}_j$ are referring to the data samples extracted from populations **x** and **y**, respectively; n and m are the number of data samples in populations x and y; and K is a kernel function such as radial basis function (RBF) kernel. A larger value of $d_{\mathrm{HSIC}}(\mathbf{x}, \mathbf{y})$ indicates that the two populations **x** and **y** are more dissimilar.

In the application of cancer response monitoring, after the computation of textural features on the parametric maps, the distance between "pre-" and "mid-treatment" scans of each patient is computed using Eq. 2. This results in a dissimilarity measure for each patient receiving cancer treatment, which can be submitted to a supervised learning paradigm for treatment response estimation as explained next.

3.3.3 Supervised Learning

Since the number of responding patients is usually much more than the number of nonresponding ones, techniques based on learning from imbalanced data [42] are needed to compensate for this uneven distribution of the data samples in the two classes. Without this compensation, the training of most of the classifiers is overwhelmed with the data samples from the majority class resulting in a very poor performance on the data samples from minority class. Random undersampling [43] has been shown to be a simple but effective approach for learning from imbalanced data in the design of a CAT system [16]. Considering $\mathcal{N}_1$ and $\mathcal{N}_2$ as the set of majority (responder) and minority (nonresponder) classes, respectively, the undersampling method randomly subsamples a set of $\mathcal{N}_1'$ from $\mathcal{N}_1$, such that $|\mathcal{N}_1'| = |\mathcal{N}_2|$.

The final stage in the design of a CAT system is to submit the balanced data to a supervised learning algorithm such as a naive Bayes classifier to train it using the ground-truth labels obtained from histological analysis. The trained classifier will be subsequently employed to predict the class label of a patient as a responder or a nonresponder. A discussion of current limitations of our techniques and future developments are provided in **Notes 8–11**.

3.4 Tissue Histology Analysis

1. For preclinical studies: Fix tumor xenografts in 10% formalin for 24–48 h. Embed the fixed specimen in paraffin. Section the specimen in five locations, aligning as much as possible with the ultrasound scan planes. We used 50 μm spacing between locations. At each location, obtain two consecutive slices—one to stain for hematoxylin and eosin (for routine histological analysis) and one to stain for TUNEL (for analysis of DNA damage).
2. For clinical studies: Following surgery, the excised and processed specimen is mounted on whole-mount 5″ by 7″ pathology slides digitized using a confocal scanner (Tissuescope, Huron Technologies). Details about the processing of surgical breast specimens can be found in Clark et al. [44]. Patients are classified as "responder" or "nonresponder" based on the tumor size change between pretreatment and end of treatment according to response evaluation criteria in solid tumors (RECIST) [1]. Tumor size is ascertained by breast magnetic resonance imaging. A tumor size reduction by 30% or more constitutes "responder" whereas a tumor size reduction by less than 30% constitutes "nonresponder," except in cases of very low residual tumor cellularity reported in the post-surgical pathology report.

4 Notes

1. We used distilled, deionized, and degassed water in all of our animal experiments (resistivity >18.0 MΩ/cm and total organic content of less than ten parts per billion). This standard is referred to as "water bath" in the text.

2. Any type of treatment able to induce apoptosis in tissue may result in changes in ultrasound images and ultrasound parameters in the treated tissue compared with the same region before treatment. However, it is recommended that the tissue chosen for investigation to have a relative homogeneous structure because ultrasound backscatter is also sensitive to the degree of randomness in scatterer arrangement. In preliminary experiments, we have measured no increase in ultrasound backscatter from cell samples with large degree of randomness in the positions of nuclei, although flow-cytometric measurements indicated that a large majority of cells (>50%) underwent death in these cell samples [45].
3. We have found that tumors larger than 10 mm (the maximum dimension) may exhibit large hyperechoic patches in ultrasound images before any treatment. Histological examination confirms that these patches are spontaneous regions of cell death. We consider that such tumors with large necrotic core cannot be used in the experiments as they may bias results. However, we have been able to utilize tumors that exhibited smaller hyperechoic patches (0.5 mm^2). This pattern of small necrotic/apoptotic regions inside the tumor prior to radiation exposure (or any other type of treatment) mimics well some human tumors and hence, we have considered that it is a valid approach to evaluate those tumors in analyses. After radiotherapy, the size of these patches typically increases covering larger tumor regions.
4. If the goal of the study is assessment of tumor cell death response to therapy, the tumor should be in the same orientation at every imaging session. The excised tumor should be sectioned in the same orientation as the ultrasound plane.
5. Some readers may prefer to use an ultrasound gel instead of water as the coupling medium. Care should be taken when using a gel to eliminate any trapped bubbles by centrifugation.
6. A spatial map of local QUS parameter estimates can be generated by assigning each sub-ROI a color-coded pixel. This is referred to as a parametric image. At high ultrasound resolutions (i.e., center frequencies above 20 MHz), parametric images of IBS have been demonstrated to distinguish areas of cell death, as presented in Fig. 7.
7. In order to obtain an accurate estimate of the effective scatterer diameter (ASD parameter; SS is also proportional to this), one must account for losses in signal intensity due to tissue attenuation. For animal tumors, attenuation of the skin and tumor was compensated for using attenuation coefficient values of 2 dB/cm-MHz and 0.6 dB/cm-MHz [28]. For human LABC tumors, an attenuation coefficient of 1 dB/cm-MHz was assumed [29].

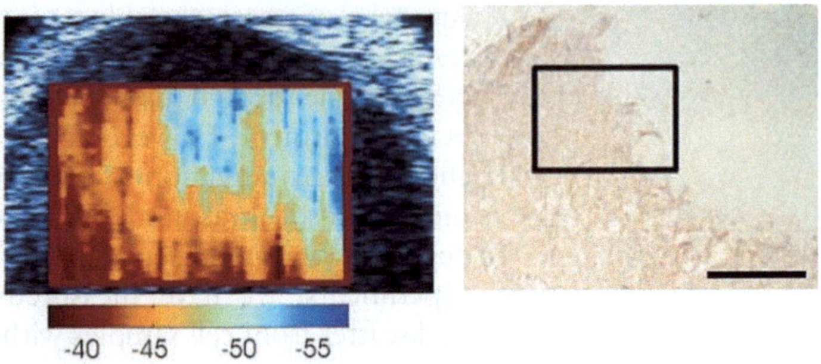

Fig. 7 Spatial map computed from the local estimates of IBS (similar to MBF) superimposed on the B-mode image and corresponding TUNEL staining. The TUNEL-stained image presents an area of cell death of similar shape as the parametric image computed from the local estimates of the ultrasound-integrated backscatter. The frame in the image of the histology corresponds approximately to the spatial map superimposed on the ultrasound image. The *color-bar* indicates the range of the corresponding estimates. The scale bar represents 1 mm. This figure is reprinted with permission from: DNA Damage Detection In Situ, Ex Vivo and In Vivo: Methods and Protocols edited by Vladimir V. Didenko (Methods in molecular biology; 682), 2011 [23]

8. In ultrasound imaging, the transmit center frequency of the ultrasound transducer determines the structures within the tissue that will scatter ultrasound. Sound at higher frequency scatters from smaller structures. Ultrasound operating at 20 MHz center frequency (77 μm) approaches cell size and therefore is sensitive to changes in cellular and nuclear structure (which occur during apoptosis). However, higher frequency sound waves are more strongly attenuated (attenuation increases with frequency in tissues), and therefore applications are limited to small animals and superficial tumors/tissues. A recent study by our group has demonstrated that conventional-frequency (~7 MHz) ultrasound BSC parameters such as AAC increase in regions of tumor cell death in the same way high-frequency ultrasound does (Fig. 3). Furthermore, image texture analysis methods examining the second order statistical properties of ultrasound parametric images have recently been demonstrated to provide a better prediction of clinical tumor response to therapy compared to the mean of ultrasound parametric maps [15].
9. Although some of the changes in spectral parameters estimates can be interpreted in terms of changes in nuclear size and acoustic impedance, recent theoretical and experimental evidence indicates that increases in nuclear randomization may in addition influence significantly the magnitude of ultrasound backscatter [45]. In these circumstances an understanding of the relative contribution of each of these nuclear changes (acoustic impendence versus randomization) to ultrasound scattering is essential in order to accurately quantify cell death.

Ongoing studies in our laboratory aim to precisely and reliably measure acoustic properties as a function of treatment [46].

10. Whereas 3D ultrasound image acquisition option is available in most small-animal ultrasound imaging systems including the one used here, such data is difficult to obtain in a clinical setting, especially with breast imaging. Whereas small animal imaging systems use mechanically actuated transducers which move across the object during acquisition, clinical imaging systems are conventionally operated by hand-held transducers, providing by default, a series of arbitrarily oriented 2D images. 3D reconstruction from hand-held ultrasound scanners is difficult due to the irregular spacing and orientation of the image planes. Reconstruction algorithms have been proposed with limited success. An alternative solution is the use of 3D ultrasound imaging systems, such as ultrasound computed tomography and rotating concave transducer technologies which have recently emerged in the market. These systems are equipped with RF data acquisition, permitting one to view ultrasound parametric images in a 3D Cartesian volume. This can provide a better visualization of response/cell death in tumors (Fig. 8). These technologies provide an operator-independent volume of the uncompressed breast (as opposed to free-hand transducers that cause breast compression during imaging), facilitating registration with other image modalities such as dynamic contrast enhanced magnetic resonance image (DCE-MRI) volume of the breast.

11. In conclusion, quantitative ultrasound spectroscopy provides safe, accurate, and noninvasive measure of DNA damage related to apoptosis in vivo, both in preclinical and clinical settings. In preclinical applications involving small animals, high-frequency ultrasound (above 20 MHz center frequency) or conventional-frequency (below 10 MHz) ultrasound can be used. In clinical applications, due to depth requirements, conventional frequency ultrasound is recommended to be used. High-frequency ultrasound parametric images provide insight into changes in cellular structure. Although conventional frequency ultrasound parametric images may not resolve very small areas of cell death (due to the lower resolution), bulk changes in cell death (greater than 0.5 mm by 0.5 mm) can be detected within 24 h in mouse models and within 1–4 weeks in patients. Quantitative ultrasound spectroscopy complements conventional ultrasound by providing more telling quantitative images for purposes of cancer diagnosis and therapy assessment, facilitating personalized medicine. With real-time imaging and therapy assessment capability, the ultrasound methods described here can be used in preclinical studies to assess responses to new experimental anticancer therapies in animal cancer models.

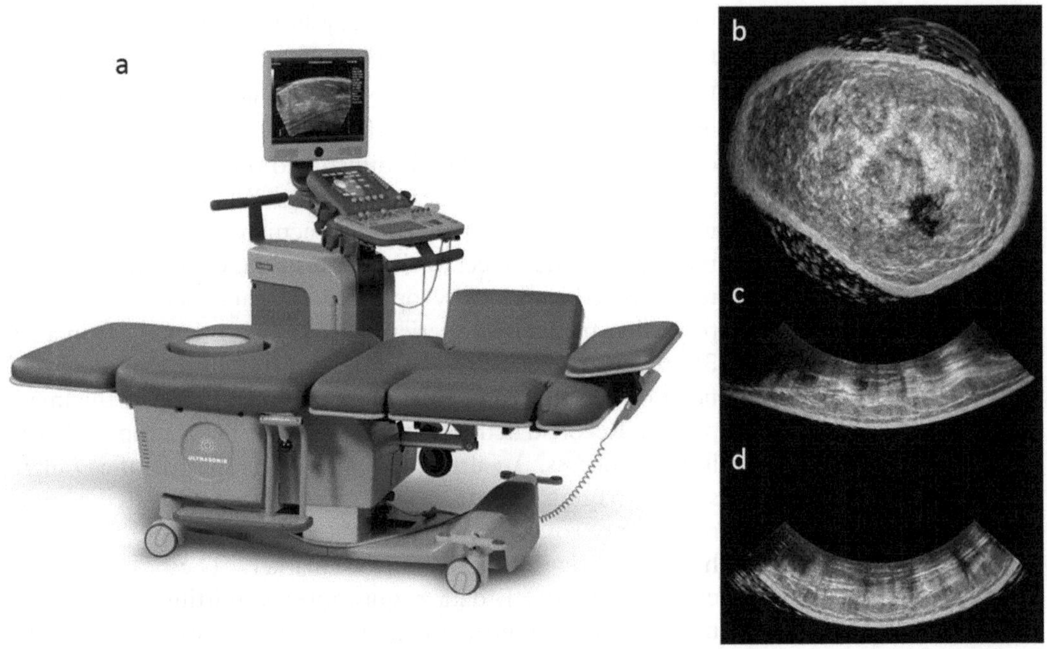

Fig. 8 Using a patented 360 ° rotating concave ultrasound transducer, Sonix Embrace Research captures realistic, uncompressed images of a breast, while the patient lies in a comfortable, prone position. The system captures gigabytes of raw data, which is ideal for cancer detection research and treatment monitoring. (**a**) Imaging system consisting of a bed with concave transducer in the center and a portable pc which handles the data acquisitions. (**b**) LR-SI view of the reconstructed breast, (**c**) AP-LR view, and (**d**) AP-SI view. Photo credits: Analogic Ultrasound (Ultrasonix, Vancouver, BC). This figure is reprinted with permission from http://www.ultrasonix.com/wikisonix/index.php/SonixEmbrace

Acknowledgments

M.J.G holds the Natural Sciences and Engineering Research Council of Canada Post-doctoral Fellowship. G.J.C. holds a University of Toronto James and Mary Davie Chair in Breast Cancer Imaging and Ablation. Funding for these projects was provided by the Terry Fox Foundation.

References

1. Eisenhauer EA, Therasse P, Bogaerts J et al (2009) New response evaluation criteria in solid tumours: revised RECIST guideline (version 1.1). Eur J Cancer 45:228–247
2. Brindle K (2008) New approaches for imaging tumour responses to treatment. Nat Rev Cancer 8:94–107
3. Chang J, Ormerod M, Powles TJ, Allred DC, Ashley SE, Dowsett M (2000) Apoptosis and proliferation as predictors of chemotherapy response in patients with breast carcinoma. Cancer 89:2145–2152
4. Nishimura R, Osako T, Okumura Y, Hayashi M, Totoyozumi Y, Arima N (2010) Ki-67 as a prognostic marker according to breast cancer subtype and a predictor of recurrence time in primary breast cancer. Exp Ther Med 1:747–754

5. Sharma U, Danishad KKA, Seenu V, Jagannathan NR (2009) Longitudinal study of the assessment by MRI and diffusion-weighted imaging of tumor response in patients with locally advanced breast cancer undergoing neoadjuvant chemotherapy. NMR Biomed 22:104–113
6. Schelling M, Avril N, Nährig J et al (2000) Positron emission tomography using [(18)F] Fluorodeoxyglucose for monitoring primary chemotherapy in breast cancer. J Clin Oncol Off J Am Soc Clin Oncol 18:1689–1695
7. Falou O, Soliman H, Sadeghi-Naini A et al (2012) Diffuse optical spectroscopy evaluation of treatment response in women with locally advanced breast cancer receiving Neoadjuvant chemotherapy. Transl Oncol 5:238–246
8. Oelze ML, O'Brien WD, Blue JP, Zachary JF (2004) Differentiation and characterization of rat mammary fibroadenomas and 4T1 mouse carcinomas using quantitative ultrasound imaging. IEEE Trans Med Imaging 23: 764–771
9. Mamou J, Coron A, Oelze M et al (2011) Three-dimensional high-frequency backscatter and envelope quantification of cancerous human lymph nodes. Ultrasound Med Biol 37(3):345–357
10. Sannachi L, Tadayyon H, Sadeghi-Naini A, Tran W, Gandhi S, Wright F, Oelze M, Czarnota G (2014) Non-invasive evaluation of breast cancer response to chemotherapy using quantitative ultrasonic backscatter parameters. Med Image Anal 20:224–236
11. Sadeghi-Naini A, Falou O, Tadayyon H, Al-Mahrouki A, Tran W, Papanicolau N, Kolios MC, Czarnota GJ (2013) Conventional frequency ultrasonic biomarkers of cancer treatment response in vivo. Transl Oncol 6:234–243
12. Sadeghi-Naini A, Papanicolau N, Falou O et al (2013) Low-frequency quantitative ultrasound imaging of cell death in vivo. Med Phys 40:82901
13. Tadayyon H, Sannachi L, Sadeghi-Naini A, Al-Mahrouki A, Tran W, Kolios MC, Czarnota GJ. Quantification of Ultrasonic Scattering Properties of In Vivo Tumor Cell Death in Mouse Models of Breast Cancer. Transl Oncol. 2015 Dec;8(6):463–73
14. Sadeghi-Naini A, Papanicolau N, Falou O et al (2013) Quantitative ultrasound evaluation of tumor cell death response in locally advanced breast cancer patients receiving chemotherapy. Clin Cancer Res 19:2163–2174
15. Sadeghi-Naini A, Sannachi L, Pritchard K, Trudeau M, Gandhi S, Wright FC, Zubovits J, Yaffe MJ, Kolios MC, Czarnota GJ (2014) Early prediction of therapy responses and outcomes in breast cancer patients using quantitative ultrasound spectral texture. Oncotarget 5:3497–3511
16. Gangeh MJ, Tadayyon H, Sannachi L, Sadeghi-Naini A, Tran WT, Czarnota GJ (2016) Computer aided Theragnosis using quantitative ultrasound spectroscopy and maximum mean discrepancy in locally advanced breast cancer. IEEE Trans Med Imaging 35: 778–790
17. Tadayyon H, Sannachi L, Gangeh M, Sadeghi-Naini A, Tran W, Trudeau ME, Pritchard K, Ghandi S, Verma S, Czarnota GJ (2016) Quantitative ultrasound assessment of breast tumor response to chemotherapy using a multi-parameter approach. Oncotarget 7:45094–45111
18. Anderson JJ, Herd M-T, King MR et al (2010) Interlaboratory comparison of backscatter coefficient estimates for tissue-mimicking phantoms. Ultrason Imaging 32:48–64
19. Kolios MC, Czarnota GJ, Lee M, Hunt JW, Sherar MD (2002) Ultrasonic spectral parameter characterization of apoptosis. Ultrasound Med Biol 28:589–597
20. Czarnota GJ, Kolios MC, Hunt JW, Sherar MD (2002) Ultrasound imaging of apoptosis. DNA-damage effects visualized. Methods Mol Biol 203:257–277
21. Brand S, Solanki B, Foster DB, Czarnota GJ, Kolios MC (2009) Monitoring of cell death in epithelial cells using high frequency ultrasound spectroscopy. Ultrasound Med Biol 35: 482–493
22. Giordano SH (2003) Update on locally advanced breast cancer. Oncologist 8:521–530
23. Didenko VV (2011) DNA damage detection in situ, ex vivo, and in vivo: methods and protocols. In: Methods in molecular biology. Springer, Berlin
24. Falou O (2011) Modelling high frequency ultrasound scattering from cells and ultrasound contrast agents. Ph.D., Ryerson University, Toronto
25. Yao LX, Zagzebski JA, Madsen EL (1990) Backscatter coefficient measurements using a reference phantom to extract depth-dependent instrumentation factors. Ultrason Imaging 12:58–70
26. Faran JJ (1951) Sound scattering by cylinders and spheres. J Acoust Soc Am 23:405–418
27. Insana MF, Wagner RF, Brown DG, Hall TJ (1990) Describing small-scale structure in random media using pulse-echo ultrasound. J Acoust Soc Am 87:179–192

28. Vlad RM, Brand S, Giles A, Kolios MC, Czarnota GJ (2009) Quantitative ultrasound characterization of responses to radiotherapy in cancer mouse models. Clin Cancer Res 15:2067–2075
29. Duric N, Littrup P, Babkin A et al (2005) Development of ultrasound tomography for breast imaging: technical assessment. Med Phys 32:1375
30. Rice SD, Heinzman JM, Brower SL, Ervin PR, Song N, Shen K, Wang D (2010) Analysis of chemotherapeutic response heterogeneity and drug clustering based on mechanism of action using an in vitro assay. Anticancer Res 30:2805–2811
31. Gangeh MJ, Sadeghi-Naini A, Kamel MS, Czarnota GJ (2013) Assessment of cancer therapy effects using texton-based characterization of quantitative ultrasound parametric images. In: 2013 IEEE 10th Int. Symp. Biomed. Imaging ISBI, pp 1372–1375
32. Ojala T, Pietikainen M, Maenpaa T (2002) Multiresolution gray-scale and rotation invariant texture classification with local binary patterns. IEEE Trans Pattern Anal Mach Intell 24:971–987
33. Varma M, Zisserman A (2005) A statistical approach to texture classification from single images. Int J Comput Vis 62:61–81
34. Varma M, Zisserman A (2009) A statistical approach to material classification using image patch exemplars. IEEE Trans Pattern Anal Mach Intell 31:2032–2047
35. Gangeh MJ, El Kaffas A, Hashim A, Giles A, Czarnota GJ (2015) Advanced machine learning and textural methods in monitoring cell death using quantitative ultrasound spectroscopy. In: 2015 IEEE 12th Int. Symp. Biomed. Imaging ISBI, pp 646–650
36. Czarnota GJ, Kolios MC, Abraham J et al (1999) Ultrasound imaging of apoptosis: high-resolution noninvasive monitoring of programmed cell death in vitro, in situ and in vivo. Br J Cancer 819(3):520–527
37. Czarnota GJ, Kolios MC (2010) Ultrasound detection of cell death. Imaging Med 2:17–28
38. Gretton A, Borgwardt K, Rasch M, Schölkopf B, Smola A (2012) A kernel two-sample test. J Mach Learn Res 13:723–773
39. Gretton A, Herbrich R, Smola A, Bousquet O, Schölkopf B (2005) Kernel methods for measuring independence. J Mach Learn Res 6:2075–2129
40. Gangeh MJ, Sadeghi-Naini A, Diu M, Tadayyon H, Kamel MS, Czarnota GJ (2014) Categorizing extent of tumor cell death response to cancer therapy using quantitative ultrasound spectroscopy and maximum mean discrepancy. IEEE Trans Med Imaging 33:1390–1400
41. Gangeh MJ, Hashim A, Giles A, Czarnota GJ (2014) Cancer therapy prognosis using quantitative ultrasound spectroscopy and a kernel-based metric. Proc SPIE 9034:903406
42. He H, Garcia EA (2009) Learning from imbalanced data. IEEE Trans Knowl Data Eng 21:1263–1284
43. Estabrooks A, Jo T, Japkowicz N (2004) A multiple resampling method for learning from imbalanced data sets. Comput Intell 20:18–36
44. Clarke GM, Eidt S, Sun L, Mawdsley G, Zubovits JT, Yaffe MJ (2007) Whole-specimen histopathology: a method to produce whole-mount breast serial sections for 3-D digital histopathology imaging. Histopathology 50: 232–242
45. Vlad R, Orlova V, Hunt J, Kolios M, Czarnota G (2008) Changes measured in the backscatter ultrasound signals during cell death can be potentially explained by an increase in cell size variance. Ultrason Imaging 29:256
46. Brand S, Weiss EC, Lemor RM, Kolios MC (2008) High frequency ultrasound tissue characterization and acoustic microscopy of intracellular changes. Ultrasound Med Biol 34: 1396–1407

Chapter 5

Fluorochrome-Labeled Inhibitors of Caspases: Expedient In Vitro and In Vivo Markers of Apoptotic Cells for Rapid Cytometric Analysis

Zbigniew Darzynkiewicz, Hong Zhao, H. Dorota Halicka, Piotr Pozarowski, and Brian Lee

Abstract

Activation of caspases is a characteristic event of apoptosis. Various cytometric methods distinguishing this event have been developed to serve as specific apoptotic markers for the assessment of apoptotic frequency within different cell populations. The method described in this chapter utilizes *fl*uorochrome *l*abeled *i*nhibitors of *ca*spases (FLICA) and is applicable to fluorescence microscopy, flow- and imaging-cytometry as well as to confocal imaging. Cell-permeant FLICA reagents tagged with carboxyfluorescein or sulforhodamine, when applied to live cells in vitro or in vivo, exclusively label the cells that are undergoing apoptosis. The FLICA labeling methodology is rapid, simple, robust, and can be combined with other markers of cell death for multiplexed analysis. Examples are presented on FLICA use in combination with a vital stain (propidium iodide), detection of the loss of mitochondrial electrochemical potential, and exposure of phosphatidylserine on the outer surface of plasma cell membrane using Annexin V fluorochrome conjugates. FLICA staining followed by cell fixation and stoichiometric staining of cellular DNA demonstrate that FLICA binding can be correlated with the concurrent analysis of DNA ploidy, cell cycle phase, DNA fragmentation, and other apoptotic events whose detection requires cell permeabilization. The "time window" for the detection of apoptosis with FLICA is wider compared to the Annexin V binding, making FLICA a preferable marker for the detection of early phase apoptosis and therefore more accurate for quantification of apoptotic cells. Unlike many other biomarkers of apoptotic cells, FLICAs can be used to detect apoptosis ex vivo and in vivo in different tissues.

Key words Apoptosis, Flow cytometry, Laser scanning cytometry, Cell death, Mitochondrial potential, Annexin V binding, Cell necrobiology

1 Introduction

Cysteine-*as*partic acid *s*pecific *prote*ases (caspases) that are being activated in response to a variety of cell death inducing stimuli, serve as critical gears within the programmed cell death machinery [1–4]. Their activation initiates specific cleavage of the respective target proteins and therefore is considered to be a marker of the

Vladimir V. Didenko (ed.), *Fast Detection of DNA Damage: Methods and Protocols*, Methods in Molecular Biology, vol. 1644,
DOI 10.1007/978-1-4939-7187-9_5, © Springer Science+Business Media LLC 2017

irreversible steps leading to cell demise. Although there are exceptions [4, 5], caspase activation is regarded to be a characteristic event of apoptosis and identification of apoptotic cells often relies on the detection of their activated state [3–6]. Caspases specifically recognize a four-amino acid sequence on their substrate proteins; the carboxyl end of aspartic acid within this sequence is their target for cleavage. Several approaches have been developed to detect the process of their activation. Because the activation involves the trans-catalytic cleavage of the zymogen pro-caspases [1], the cleavage products having lower molecular weight than the zymogen can be revealed electrophoretically and identified in Western blots using caspase-specific antibodies. Chromogenic/fluorogenic modifications of the four-amino-acid, preferred caspase targeting sequences led to the creation of caspase detection substrates. These caspases substrates, which initially are colorless or non-fluorescent upon cleavage by caspases, generate a colored or fluorescing signal. These substrates were developed for the purpose of detecting caspases activation. Early on, their utility was confined to the assessment of the presence of activated caspases in cell extracts. When these chromogenic/fluorogenic caspase detection substrates were chemically modified to become cell membrane permeant, this modification allowed for the detection of caspases activation in whole living cells. Elimination of a cell-lysis requirement leads to several advantages: (1) creation of a whole-cell-based, flow cytometry compatible, caspase detection assay format, [7, 8] and (2) elimination of the possibility for unintended artifact creation resulting from cell lysis treatments. Adapted to cytometry is also the methodology based on fluorescence resonance energy transfer (FRET) that detects the cleavage of the caspase substrate that contains the FRET-responsive fluorophores [9, 10]. Immunocytochemical detection either of the activated (trans-catalytically cleaved) caspase [6, 11] or of the caspase specific cleavage product, such as the p89 fragment of poly(ADP-ribose) [12], are other methodologies used to detect activity of these proteolytic enzymes in single cells, applicable to cytometry.

The method described here relies on the use of the *fl*uorochrome-*l*abeled *i*nhibitors of *ca*spases (FLICA) [13–15]. The methodology is based on the use of enzyme-specific inhibitors tagged with fluorochromes such as carboxyfluorescein (FAM)- or fluorescein (FITC)-peptide-fluoromethyl ketone (FMK), the ligands that bind to the active centers of caspases. These FMK reagents penetrate through the plasma membrane of live cells and are nontoxic to the cell. In fact, in some cell systems by virtue of caspase inhibition, FLICAs slow down the process of apoptosis and prevent total cell disintegration. As a consequence of their caspase inhibition properties, the use of FLICA probes was proposed as an accurate dynamic (real time) assay of apoptosis for different cell populations [16]. The recognition peptide moiety of these reagents was expected to provide some level of specificity of their binding to particular caspases. Thus, the FLICAs containing VDVAD, DEVD, VEID, YVAD, LETD, LEHD, and AEVD peptide moieties were expected to preferentially bind to activated caspases-2, −3, −6, −1, −8, −9, and −10, respectively. On the other hand, the FAM-VAD-FMK containing the valyl-alanyl-aspartic acid residue sequence was designed to be pan-caspase reactive, binding to activated caspases-1, −3 -4, −5, −7, −8, and −9 [17]. It was later observed, however, that because of the FMK group reactivity, likely with thiol groups of intracellular proteins that become available upon cleavage by caspases, their binding is not as caspase-specific as initially thought [15]. Despite the apparent lack of specificity in labeling individual caspases, the FLICA probes have consistently shown themselves to be highly reliable reporters of caspases activation and convenient markers of apoptotic cells. In contrast to a variety of other markers of apoptotic cells (e.g., annexin V) FLICA, also named FLIVO™, can be used to rapidly detect the caspase-driven apoptosis events upon proper, fluorochrome-specific wavelength excitation of live tissues or in vivo following their injection into animals [18, 19].

Exposure of live cells to FLICAs results in the rapid uptake of these reagents followed by their covalent binding to apoptotic cells containing activated caspase enzymes. Unbound FLICAs are removed from the non-apoptotic cells via a simple concentration gradient diffusion process following repeated rinsing of the cells with wash-buffer. The protocols given below describe the labeling of cells containing activated caspases using FAM-VAD-FMK. The same protocol can be applied to other FLICAs. Concurrent exposure of cells to propidium iodide (PI) strongly labels all cells with a compromised plasma membrane integrity that cannot exclude this cationic dye; the loss of membrane integrity is a key feature of mid-late stage apoptotic cells or cells undergoing necrosis. On the other hand, simultaneous cell staining with the mitochondrial electrochemical potential probe, chloromethyl-X-rosamine (CMXRos; MitoTracker Red) [20] and FLICA allows one to discriminate between two sequential events of apoptosis: dissipation of the inner mitochondrial membrane potential and activation of the caspase enzyme cascade, respectively. Likewise, simultaneous staining with FLICA and Annexin V conjugated to red fluorescing dyes such as Cy5 reveals the relationship between caspases activation and exposure of phosphatidylserine (PS) residues on the external surface of plasma membrane during apoptosis [7, 21]. Cells labeled with FLICAs and PI, CMXRos or Annexin V-Cy5 can be examined by fluorescence microscopy or subjected to quantitative analysis by flow- or image- cytometry such as laser scanning cytometry (LSC). By virtue of the ability to rapidly measure large cell populations and to analyze cell images, LSC is particularly useful in studies of apoptosis [22, 23].

2 Materials

1. Cells to be analyzed: Can be grown on slides (*see* Subheading 3.1) or in suspension.
2. Microscope slides, coverslips, and culture vessels having microscope slides at the bottom of the chamber, such as Chamberslide", Nunc, Inc., or single- or multi-chambered Falcon Culture Slides (BD Biosciences): *see* Subheading 3.1.
3. Instrumentation: Fluorescence microscope, or multiparameter Laser Scanning Cytometer (LSC; iCys^{R} model), available from Thorlabs Inc. (Newton, NJ, USA). Flow cytometers of different types, offered by several manufacturers, can be used to measure cell fluorescence following staining according to the procedures described below. The manufacturers of the most common flow cytometers are Beckman /Coulter Corporation (Miami, FL), BD Biosciences (formerly Becton Dickinson Immunocytometry Systems; San Jose, CA), DACO/Cytomation (Fort Collins, CO), and PARTEC GmbH (Zurich, Switzerland).

   The software to deconvolve the DNA content frequency histograms, to analyze the cell cycle distributions, is available from Phoenix Flow Systems San Diego, CA) or Verity Software House (Topham, MA).
4. Phosphate-buffered saline (PBS).
5. Dimethylsulfoxide (DMSO).
6. Stock solution of PI: Dissolve 1 mg of PI in 1 mL of distilled water. This solution can be stored at 4 °C in the dark for several months.
7. Stock FLICA solution: Dissolve lyophilized FLICA (e.g., FAM-VAD-FMK; available as a component of the FLICA™ Apoptosis Detection ("Green FLICA") kit from Immunochemistry Technologies LLC, Bloomington, MN) in dimethylsulfoxide (DMSO) as specified in the kit to obtain 150× concentrated (stock) solution of this inhibitor. Also available from this vendor are caspase-2 (VDVAD), caspase-3 (DEVD), caspase-6 (VEID), caspase-1 (YVAD), caspase-8 (LETD), caspase-9 (LEHD), and caspase-10 (AEVD) FLICAs. Aliquots of FLICAs solution (150× in DMSO) may be stored at −20 ° C in the dark for 6 months.
8. Intermediate (30× concentrated) FLICA solution: Prepare a 30-× concentrated solution of FAM-VAD-FMK by diluting the stock solution 1:5 in PBS. Mix the vial until the contents become transparent and homogenous. This solution should be made freshly from the 150× DMSO stock solution just prior to adding it to the cells. Protect all FLICA solutions from light.

9. FLICA staining solution: just prior to the use, add 3 μL of 30× concentrated FAM-VAD-FMK solution into 100 μL of culture medium.
10. Rinsing solution: 1% (w/v) BSA in PBS.
11. Staining solution of PI: Add 10 μL of stock solution of PI to 1 mL of the rinsing solution.
12. Mito-Tracker Red (Invitrogen/Molecular Probes).
13. Annexin V-Cy5 conjugate and the binding buffer (Abcam, Cambridge, MA or Alexis Biochemicals, San Diego, CA).
14. Parafilm "M."

3 Methods

3.1 Attachment of Cells to Slides (Cells to be Analyzed by Microscopy and/or LSC)

The procedure requires incubation of live (unfixed, not permeabilized) cells with solutions of FLICAs. A variety of adherent cells are available for growth in cell culture flasks. Such cells can be attached to microscope slides by culturing them on slides or coverslips. Culture vessels having microscope slides at the bottom of the chamber are commercially available (*see* Subheading 2, **item 2**). The cells growing in these chambers spread and attach to the slide surface after incubation at 37 °C for several hours. Alternatively, the cells can be grown on coverslips, e.g., placed on the bottom of Petri dishes. The coverslips are then inverted over shallow (<1 mm) wells on the microscope slides. The wells can be prepared by constructing the well walls (~2 × 1 cm size) either with a pen that deposits a hydrophobic barrier ("Isolator", Shandon Scientific), nail polish, or melted paraffin. The wells also may be made by preparing a strip of Parafilm "M" of the size of the slide, cutting a hole ~2 × 1 cm in the middle of this strip, placing the strip on the microscope slide and heating the slide on a warm plate until the Parafilm starts to melt. It should be stressed, however, that because the cells detach during late stages of apoptosis [6, 23] these apoptotic cells may be selectively lost if the analysis is limited to attached cells.

Cells that normally grow in suspension can be attached to glass slides by electrostatic forces. This is due to the fact that sialic acid residues that cover the cell surface have a net negative charge in contrast to the glass surface which is positively charged. Incubation of cells on microscope slides in the absence of any serum or serum proteins (which otherwise neutralize the charge), thus, leads to their attachment. The cells taken from culture should be rinsed in PBS in order to remove serum proteins contained in the cell culture media and then resuspended in PBS at a concentration of 2×10^5 cells/mL. An aliquot (50–100 μL) of this suspension should be deposited within a shallow well (prepared as described above) on the horizontally placed microscope slide. To prevent drying, a small piece (~2 × 2 cm) of a thin polyethylene foil or Parafilm may be placed atop of the cell suspension drop. A short (15–20 min) incubation of such cell suspension at room temperature in a closed box containing wet tissue or filter paper that provides 100% humidity is adequate to ensure that most cells will firmly attach to the slide surface. Cells attached in this manner remain viable for several hours and can be subjected to surface immmunophenotyping, viability tests, or intracellular enzyme kinetics assays. Such preparations can then be fixed (e.g., in formaldehyde) without a significant loss of cells from the slide. However, as in the case of cell growth on glass [23], late apoptotic cells have a tendency to detach even after initial attachment.

It should be stressed that the microscope slide to which cells are going to be attached electrostatically should be extra clean. Fingerprints leave oils on the slide that interferes with cell attachment. To remove possible contamination of glass surface that may interfere with cell attachment, soak microscope slides in a household detergent and then rinse in water and 100% ethanol respectively. Slides should be allowed to air dry and used the same day they were cleaned.

3.2 Cell Staining and Analysis by Microscopy or LSC/(iCys[R] Research Imaging Cytometer)

1. Attach the cells to the microscope slide as described in Subheading 3.1 Keep the cells immersed in the culture medium by adding 100 μL of the medium (with 10% serum) into the well on the microscope slide to cover the area with the cells. In the case of cells growing on microscope slide chambers, move directly to **step 2**.
2. Remove the medium and replace it with 100 μL of the 1× FLICA (e.g., FAM-VAD-FMK) staining solution.
3. Place ~2 × 4 cm strip of Parafilm atop the staining solution to prevent drying. Incubate the slides horizontally for 30 or up to 60 min at 37 ° C in a closed box with wet tissue or filter paper to ensure 100% humidity, in the dark.
4. Remove the staining solution with Pasteur pipette. Rinse twice with the rinsing solution each time, adding a new aliquot, gently mixing, and after 2 min replacing with the next rinse.
5. Apply 100 μL of the PI staining solution atop of the cells deposited on the slide. Cover with a coverslip and seal the edges to prevent drying (*see* **Notes 1–4**).
6. During the subsequent 30 min after the addition of PI solution, observe the cells under a fluorescence microscope (blue light illumination) or measure cell fluorescence with LSC. Use the argon ion laser (488 nm) of LSC to excite fluorescence, contour on light scatter, and measure green fluorescence of FLICA at 530 ± 20 nm and red fluorescence of PI at >600 nm (Fig. 1) (*see* **Note 5**).

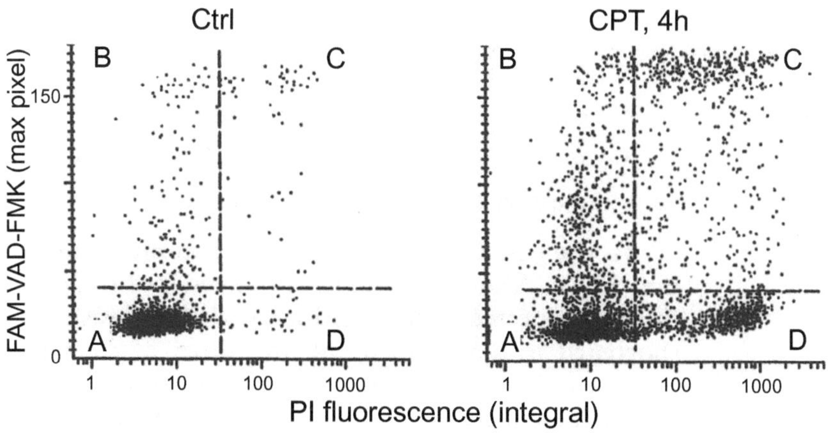

Fig. 1 The bivariate distributions (scatterplots) of FAM-VAD-FMK (FLICA; green maximal pixel) vs PI (red integral) fluorescence of the control (Ctrl) and camptothecin (CPT) treated HL-60 cells. To induce apoptosis the cells were treated for 4 h with 0.15 μM of the DNA topoisomerase I inhibitor (CPT). The cells were then stained according to the protocol (Subheading 3.2) and their fluorescence was measured by LSC (iCys[R]). The live non-apoptotic cells, which are predominant in Ctrl, are unlabeled (quadrant A). Early apoptotic cells have increased FLICA fluorescence but minimal fluorescence of PI (quadrant B). The cells more advanced in apoptosis show variable degrees of both, FLICA-positive and PI-positive, fluorescence (quadrant C). At very late stages of apoptosis ("necrotic stage"), caspases either leak out of cells or are not reactive with FLICA and cells become FLICA-negative/PI-positive (quadrant D)

When cells are attached to microscope slides, the shortest time required for their combined FLICA and PI staining is only slightly over 30 min. FLICA incubation can be minimally 30 minutes, and within moments of adding PI, the cells can be analyzed visually or by LSC.

3.3 Cell Staining and Analysis by Flow Cytometry

1. Suspend 5 × 10^5–10^6 cells in 0.3 mL of complete culture medium (with 10% serum) in a centrifuge tube.
2. Add 10 μL of the 30× concentrated ("intermediate") FLICA staining solution to this cell suspension. Mix the cell suspension by flicking the tube.
3. Incubate for 30–60 min at 37 ° C in atmosphere of air with 5% CO_2, at 100% humidity, in the dark.
4. To the cell suspension with FLICA, add 5 mL of the rinsing solution (PBS with BSA) or 1× "wash buffer" provided with the FLICA kit and gently mix the cell suspension.
5. Centrifuge at 300 × *g* for 5 min at room temperature and remove the supernatant by aspiration.
6. Resuspend cell pellet in 2 mL of the rinsing solution or in 1× "wash buffer."
7. Centrifuge at 300 × *g* for 5 min and aspirate the supernatant.

8. Resuspend cells in 1 mL of the PI staining solution. Place the tube on ice (*see* **Notes 2–4**).
9. Measure cell fluorescence by flow cytometry
   - excite cell fluorescence with 488 nm laser line.
   - measure green fluorescence of FLICA at 530 ± 20 nm.
   - measure red fluorescence of PI at >600 nm.

In the case of suspended cells the duration of their combined FLICA and PI staining can be reduced to 30 min, when the two repeated centrifugations with the "wash buffer" are shortened to 2 min each and the speed of centrifugation is increased (500 × *g*). Avoid exposing cells to g forces >800 × *g*. Under these increased *g*-force conditions, there is an increased risk of fragile, compromised cell membrane integrity (apoptotic-cell) losses due to excessive *g*-force packing and membrane structure disruption (*see* Notes 6–11).

4 Notes

1. Protect cells from light throughout the procedure.
2. Staining with PI is optional. It allows one to identify the cells that have compromised cell membrane integrity. Such cells (necrotic and mid-late stage apoptotic cells, cells with structurally compromised plasma membranes, isolated cell nuclei) will stain red as a result of their inability to exclude PI (Fig. 1).
3. *If concurrently with FLICA cells are stained with CMXRos (MitoTracker Red):* Instead of adding PI solution as described above, add 100 μL of PBS containing 0.2 μM CMXRos. Alternatively, the CMXRos potentiometric dye can be interchanged with a tetramethylrhodamine derivative dye like tetramethylrhodamine methyl ester perchlorate (TMRM, available from Sigma, Molecular Probes/ThermoFisher or in kit form from Immunochemistry Technologies). Cells can be stained with 0.4 μM of TMRM for 15 min at 37 °C. After the addition of either the CMXRos or TMRM potentiometric dyes, rinse the cells with rinsing solution, suspend in rinsing solution, and measure their green (FLICA) and red (CMXRos or TMRM) fluorescence the same way as in the case of cells stained with FLICA and PI (Fig. 2).
4. *If concurrently with FLICA cells are stained with Annexin V-Cy5 conjugate:* Instead of adding PI solution, add 100 μL of the 1× binding buffer containing Annexin V-Cy5 at the concentration suggested by the vendor. Incubate for 15 min, rinse once with the binding buffer, and suspend in rinsing buffer. The calcium containing binding buffer must be used to maintain the affinity

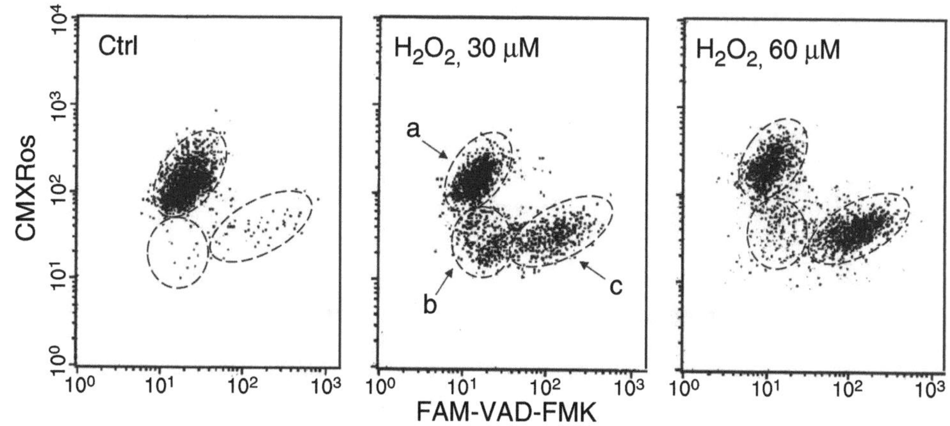

Fig. 2 Concurrent detection of caspases activation by FLICA (FAM-VAD-FMK) and the loss of mitochondrial electrochemical potential ($\Delta\psi_m$) during apoptosis. To induce apoptosis, human leukemic Jurkat cells were treated with 30 μM or 60 μM H_2O_2 for 7 h [7]. The cells were then subjected to staining with FLICA and CMXRos (MitoTracker Red) and their fluorescence was measured by flow cytometry as described in the protocol (Subheading 3.3 and **Note 3**). Based on differential binding of FLICA and CMXRos, one can distinguish three cell subpopulations: (*a*) live, non-apoptotic cells; (*b*) cells having decreased mitochondrial potential (CMXRos binding), prior to caspases activation, and (*c*) cells showing both decreased mitochondrial potential and increased caspases activation (*c*). The data illustrate that the reduction of mitochondrial potential precedes FLICA binding. Treatment with 60 μM H_2O_2 accelerates the process of apoptosis as evidenced by an increase in frequency of FLICA-positive cells. (For more details *see* Ref. 15)

interaction between the Annexin and PS residues on the apoptotic cell membrane surface. Excite cell fluorescence with both 488 nm and red diode (647 nm) lasers and measure the emission of Cy5 at far red (>650 nm) wavelength (Fig. 3). This procedure, however, can also be combined with PI staining. High-intensity PI fluorescence at red wavelength (590–620 nm, excited at 488 nm) allows one to preidentify and gate out in advance the cells with a damaged plasma membrane (PI-positive). PI-negative cells may then be analyzed with respect to their Annexin V –Cy5 and FLICA fluorescence properties.

5. After **step 4** (Subheading 3.2) or **step 7** (Subheading 3.3) the cells may be fixed in 1% formaldehyde (10 min on ice) followed by 70% ethanol and then subjected to staining with PI in the presence of RNase A or stained with 7-aminoactinomycin D. In this format, the PI is used to stain the DNA of all cells to facilitate the assessment of DNA ploidy parameters. Analysis of the FLICA vs. PI fluorescence by LSC or flow cytometry allows for the correlation of caspases activation with cellular DNA content, i.e., the cell cycle position or DNA ploidy. Details of this procedure are provided in reference [6, 24]. Alternatively, when two-laser excitation is available and one of the lasers produces UV light, the cellular DNA may be counterstained with Hoechst 33342 or with 4′,6-diamidino-2-phenylindole (DAPI), both available from Molecular Probes (ThermoFisher Scientific, Inc.).

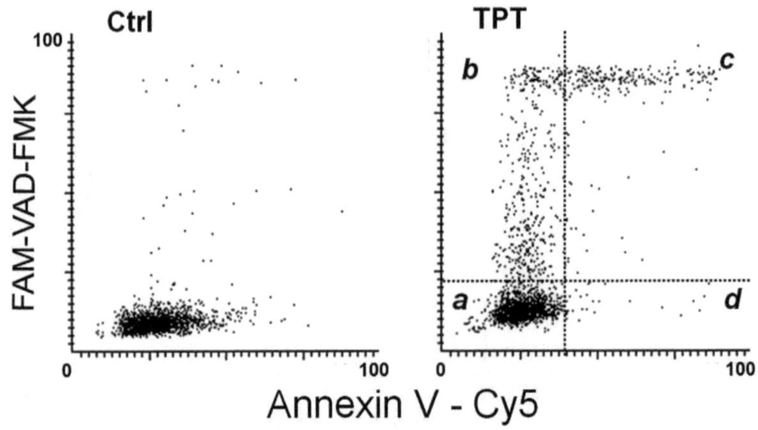

Fig. 3 Relationship between caspases activation as detected by FLICA binding and externalization of phosphatidylserine revealed by Annexin V binding during apoptosis. Apoptosis of HL-60 cells was induced by treatment with the DNA topoisomerase I inhibitor topotecan (TPT, 0.15 μM, 90 min). The cells were then labeled with FLICA (FAM-VAD-FMK), and subsequently with annexin V-Cy5 conjugate, and their fluorescence was measured by LSC as described in the protocol (Subheading 3.2 and **Note 4**). Four quadrants shown in the right panel identify cells that are (*a*) non-apoptotic, (*b*) early apoptotic cells that bind FLICA but do not bind annexin V-Cy5, (*c*) apoptotic cells that bind both FLICA and annexin V-Cy5, and (*d*) very late apoptotic or necrotic cells that are FLICA-negative. The data show that the "time window" to detect apoptosis is wider for the FLICA assay as the early apoptotic cells are FLICA-positive (*b*) and not detectable by the Annexin V-Cy5 assay

6. One has to keep in mind that FLICAs are not passive reagents that mark the activated caspases, but instead, react directly with the caspase enzyme by covalent interaction with reactive cysteine within the active catalytic site of the enzyme. This inhibits caspase activity suppressing the process of apoptosis. Thus, the rate of apoptosis progression and all the events related to caspases activity are suppressed by FLICAs [16].
7. Another problem that should be taken into account when using FLICA to mark the activated caspases in live cells pertains to the structural fragility of apoptotic cells. The flow cytometric assay requires incubation of live cells with these reagents followed by repeated rinsing to remove unbound FLICA from the non-apoptotic cells. Apoptotic cells, particularly at late stages of apoptosis, are fragile and may be preferentially lost during the centrifugations. A certain degree of stability is derived from the presence of serum (up to 20% v/v) or BSA (up to 2% w/v) in the rinsing buffers. Also, the cells should be sedimented with minimal *g*-force and short centrifugation time. Because of a possibility of a selective loss of apoptotic cells, one has to be careful drawing conclusions about the absolute frequency of apoptosis based on the percentage of FLICA-positive cells in the samples assayed by flow cytometry. In the case of cell analysis by LSC/iCys® the propensity of apoptotic cells to detach in cultures and thus escape from analysis should also be taken into account, as it may favor a downward bias in the estimation of

the apoptotic index as well. The above technical difficulties in the analysis of frequency of apoptosis pertain to any cytometric methodology, not only by FLICA and are discussed in extent elsewhere [23].

8. It is difficult to assess the specificity of in situ bound individual FLICA sequences designed to be markers for their respective caspases in light of the evidence of a nonspecific component in FLICA binding [15]. While activation of caspases is definitely associated with FLICA binding, it is likely that products of cleavage of other proteins by caspases have exposed thiol groups reactive with FLICA as well [15].
9. Also available are FLICA red fluorescing reagents containing sulforhodamine (SR) instead of carboxyfluorescein (FAM). Their availability extends multiparameter/multiplexed FLICA applications in combination with other markers.
10. A combination of FLICA with SYTO dyes offers an attractive marker of apoptosis, concurrently revealing both the activation of caspases and the condensation of nucleic acids [25]. Combining the red-fluorescing and nontoxic marker of cell membrane integrity loss DRAQ7 [26] with green fluorescing FLICAs provides an excellent arrangement of dual dynamic (real time) indicators of cell viability and apoptosis.
11. It should also be stressed that because of FLICA's nontoxicity properties at the concentrations used to label apoptotic cells, these reagents are ideally suited to serve as ex vivo and in vivo markers of apoptosis induction events in these cells. When formulated and vialed for specific use as an in vivo apoptosis detection marker, defined as FLIVO™ (Immunochemistry Technologies, LLC), these imaging reagents were successfully used to allow in vivo labeling and subsequent ex vivo detection and assessment, of apoptotic cells within a variety of rat and mouse organ tissues following apoptosis induction by way of a number of different induction routes [26–30].

Acknowledgment

Supported by the Robert A. Welke Cancer Research Foundation.

References

1. Murray J, Renslo AR (2013) Modulating caspase activity: beyond the active site. Curr Opin Struct Biol 23:812–819
2. Earnshaw WC, Martins LM, Kaufmann SH (1999) Mammalian caspases: structure, activation, substrates, and functions during apoptosis. Annu Rev Biochem 68:383–424
3. Nicholson DW (1999) Caspase structure, proteolytic substrates and function during apoptotic cell death. Cell Death Differ 6:1028–1042

4. Salvesen GS, Riedl SJ (2008) Caspase mechanisms. Adv Exp Med Biol 615:13–232
5. Lalaoui N, Lindqvist LM, Sandow JJ, Ekert PG (2015) The molecular relationships between apoptosis, autophagy and necroptosis. Semin Cell Dev Biol 39:63–69
6. Wlodkowic D, Telford W, Skommer J, Darzynkiewicz Z (2011) Apoptosis and beyond: cytometry in studies of programmed cell death. Methods Cell Biol 103:55–98
7. Komoriya A, Packard BZ, Brown MJ, Wu ML, Henkart PA (2000) Assessment of caspase activities in intact apoptotic thymocytes using cell-permeable fluorogenic caspase substrates. J Exp Med 191:1819–1828
8. Lee BW, Johnson GL, Hed SA, Darzynkiewicz Z, Talhouk JW, Mehrota S (2003) DEVDase detection in intact apoptotic cells using the cell permeant fluorogenic substrate, $(z\text{-DEVD})_2$-cresyl violet. BioTechniques 35:1080–1085
9. Alasia S, Cocito C, Merighi A, Lossi L (2015) Real-time visualization of caspase-3 activation by fluorescence resonance energy transfer (FRET). Methods Mol Biol 1254:99–113
10. Yu JQ, Liu XF, Chin LK, Liu AQ, Luo KQ (2013) Study of endothelial cell apoptosis using fluorescence resonance energy transfer (FRET) biosensor cell line with hemodynamic microfluidic chip system. Lab Chip 13:2693–2700
11. Huang X, Okafuji M, Traganos F, Luther E, Holden E, Darzynkiewicz Z (2004) Assessment of histone H2AX phosphorylation induced by DNA topoisomerase I and II inhibitors topotecan and mitoxantrone and by DNA crosslinking agent cisplatin. Cytometry A 58A:99–110
12. Li X, Darzynkiewicz Z (2000) Cleavage of poly(ADP-ribose) polymerase measured *in situ* in individual cells: relationship to DNA fragmentation and cell cycle position during apoptosis. Exp Cell Res 255:125–132
13. Bedner E, Smolewski P, Amstad P, Darzynkiewicz Z (2000) Activation of caspases measured in situ by binding of fluorochrome-labeled inhibitors of caspases (FLICA): correlation with DNA fragmentation. Exp Cell Res 260:308–313
14. Smolewski P, Bedner E, Du L, Hsieh T-C, Wu JM, Phelps DJ, Darzynkiewicz Z (2001) Detection of caspases activation by fluorochrome-labeled inhibitors: multiparameter analysis by laser scanning cytometry. Cytometry 44:73–82
15. Pozarowski P, Huang X, Halicka DH, Lee B, Johnson G, Darzynkiewicz Z (2003) Interactions of fluorochrome-labeled caspase inhibitors with apoptotic cells. A caution in data interpretation. Cytometry A 55A:50–60
16. Smolewski P, Grabarek J, Lee BW, Johnson GL, Darzynkiewicz Z (2002) Kinetics of HL 60 cell entry to apoptosis during treatment with TNF-α or camptothecin assayed by the stathmo-apoptosis method. Cytometry 47:143–149
17. Thornberry NA, Rano TA, Peterson EP, Rasper DM, Timkey T, Garcia-Calvo M, Houtzager VM, Nordstrom PA, Roy S, Valliancourt JP, Chapman KT, Nicholson DW (1997) A combinatorial approach defines specificities of members of the caspase family and granzyme B. J Biol Chem 272:17907–17911
18. Barreiro-Iglesias A, Shifman MI (2012) Use of fluorochrome-labeled inhibitors of caspases to detect neuronal apoptosis in the whole-mounted lamprey brain after spinal cord injury. Enzyme Res 2012:835–731
19. Gómez-Cabañas L, Delgado-Martín C, López-Cotarelo P, Escribano-Diaz C, Alonso-C LM, Riol-Blanco L, Rodríguez-Fernández JL (2014) Detecting apoptosis of leukocytes in mouse lymph nodes. Nat Protoc 9:1102–1112
20. Pendergrass W, Wolf N, Poot M (2004) Efficacy of MitoTracker Green and CMXRosamine to measure changes in mitochondrial membrane potentials in living cells and tissues. Cytometry A 61:162–169
21. Koopman G, Reutelingsperger CP, Kuijten GA, Keehnen RM, Pals ST, van Oers MH (1994) Annexin V for flow cytometric detection of phosphatidylderine expression of B cells undergoing apoptosis. Blood 84:1415–1420
22. Pozarowski P, Holden E, Darzynkiewicz Z (2013) Laser scanning cytometry: principles and applications. Methods Mol Biol 913:187–212
23. Darzynkiewicz Z, Bedner E, Traganos F (2001) Difficulties and pitfalls in analysis of apoptosis. Methods Cell Biol 63:527–559
24. Darzynkiewicz Z, Zhao H (2014) Analysis of cell cycle by flow cytometry. In: eLS (Encyclopedia of Life Sciences), John Wiley & Sons Ltd, Chichester
25. Wlodkowic D, Skommer J, Hillier C, Darzynkiewicz Z (2008) Multiparameter detection of apoptosis using red-excitable SYTO probes. Cytometry A 73A:563–569
26. Akagi J, Kordon M, Zhao H, Matuszek A, Dobrucki J, Errington R, Smith PJ, Takeda K, Darzynkiewicz Z, Wlodkowic D (2013) Real-time cell viability assays using a new anthracycline derivative DRAQ7. Cytometry A 83A:227–234
27. Lee BW, Olin MR, Johnson GL, Griffin RJ (2008) *In vitro* and *in vivo* apoptosis detection

using membrane permeant fluorescent-labeled inhibitors of caspases. Methods Mol Biol 414:109–135

28. Tsai YC, Mendoza A, Mariano JM, Zhou M, Kostova Z, Chen B, Veenstra T, Hewitt SM, Helman LJ, Khanna C, Weissman AM (2007) The ubiquitin ligase gp78 promotes sarcoma metastasis by targeting KAI1 for degradation. Nat Med 13:1504–1509

29. Cursio R, Colosetti P, Auberger P, Gugenheim J (2008) Liver apoptosis following normothermic ischemia-reperfusion: in vivo evaluation of caspase activity by FLIVO assay in rats. Transplant Proceed 40:2038–2041

30. Barreiro-Iglesias A, Shifman MI (2015) Detection of activated caspase-8 in injured spinal axons by using fluorochrome-labeled inhibitors of caspases (FLICA). Methods Mol Biol 1254:329–339

Chapter 6

The Fast-Halo Assay for the Detection of DNA Damage

Piero Sestili, Cinzia Calcabrini, Anna Rita Diaz, Carmela Fimognari, and Vilberto Stocchi

Abstract

The need for express screening of the DNA damaging potential of chemicals has progressively increased over the past 20 years due to the wide number of new synthetic molecules to be evaluated, as well as the adoption of more stringent chemical regulations such as the EU REACH and risk reduction politics. In this regard, DNA diffusion assays such as the microelectrophoretic comet assay paved the way for rapid genotoxicity testing. A more significant simplification and speeding up of the experimental processes was achieved with the fast halo assay (FHA) described in the present chapter. FHA operates at the single cell level and relies on radial dispersion of the fragments of damaged DNA from intact nuclear DNA. The fragmented DNA is separated by diffusion in an alkaline solvent and is stained, visualized, and finally quantified using computer-assisted image analysis programs. This permits the rapid assessment of the extent of DNA breakage caused by different types of DNA lesions. FHA has proven to be sensitive, reliable, and flexible. This is currently one of the simplest, cheapest, and quickest assays for studying DNA damage and repair in living cells. It does not need expensive reagents or electrophoretic equipment and requires only 40 min to prepare samples for computer-based quantification. This technique can be particularly useful in rapid genotoxicity assessments and in high-throughput genotoxicity screenings.

Key words DNA damage, DNA repair, Single-strand breaks, Double-strand breaks, Apoptosis, Large-scale genotoxicity screening

1 Introduction

A variety of chemicals and physical agents can damage cellular DNA through specific lesions such as single- and double-strand breaks (SSBs and DSBs, respectively), modified bases, DNA–DNA crosslinks, and DNA–protein crosslinks [1]. These lesions may trigger an array of biological responses from the cellular to the organ level, one of the most dangerous being carcinogenesis [1]. Hence, the detection of these lesions and the identification of DNA-damaging agents/chemicals are of the utmost importance from both biological and toxicological points of view.

Advances in understanding of the negative interactions of genotoxic agents with the structure and function of the genome also

Vladimir V. Didenko (ed.), *Fast Detection of DNA Damage: Methods and Protocols*, Methods in Molecular Biology, vol. 1644, DOI 10.1007/978-1-4939-7187-9_6, © Springer Science+Business Media LLC 2017

depend on the development and availability of more robust, flexible, and reliable techniques. During the 1970s, a variety of methods were developed for the direct and indirect detection of DNA damage and repair in eukaryotic cells. These included cesium chloride equilibrium density gradients, benzoylated naphthoylated diethylaminoethyl cellulose columns, 313 nm irradiation of DNA-containing bromodeoxyuridine, alkaline filter elution, alkaline sucrose gradients, and direct assays for the identification of adducts from DNA [2–5]. Although fundamental for the development of genetic toxicology, these techniques were cumbersome, expensive, time consuming, and inadequate for large-scale screening purposes. During the 1980s a new and independent technique, the comet assay, also known as single cell microelectrophoresis, was developed [6]. The comet assay was the first technique capable of combining the simplicity of biochemical techniques for detecting DNA damage with the single cell approach typical of cytogenetic assays.

Advantages of the comet assay include: (1) its relative simplicity as compared to older methods; (2) the need for small numbers of cells per sample (<10,000); (3) its flexibility (any eukaryotic cell type can be analyzed); (4) no need for radioactive labeling of DNA; (5) and its sensitivity [7]. Thanks to these features, the comet assay is recognized by regulatory agencies as an indicator test for large-scale screenings.

In search of alternative, faster, and simpler high-throughput, early prescreening methods to assess the DNA-damaging potential of chemical agents, our group developed two "radial diffusion" assays: the alkaline halo assay (AHA) in 1999 [8] and the subsequent fast halo assay (FHA) in 2006 [9].

These assays have been used by our and other groups to analyze DNA strand breakage induced by a number of agents or apoptotic cell death [8–58].

The fast halo assay was named so [8] by virtue of its unique rapidity and simplicity as compared to its precursor AHA, comet assay, or other methods to detect nuclear DNA lesions. FHA is based on the observation that fragmented DNA from damaged cells can be radially separated from intact chromatin with a simple incubation in an alkaline solution, i.e., without the need for either the long lysis stage of AHA and comet assay or for the electrophoretic process that characterizes the comet assay. Omitting these stages reduces both the total experimental time and the expense to set up the assay.

In FHA, the diffusion process is promoted by alkaline—be it denaturing or not—solvation of nuclei which produces, upon DNA staining and microscopic examination, radial and centrifugal migration of fragmented DNA which forms circular "halos" concentric to the smaller nuclear remnants. The size of the halos is directly related to the number/frequency of the DNA lesions, i.e., the bigger the halo the higher the damage (*see* Figs. 1 and 2), as

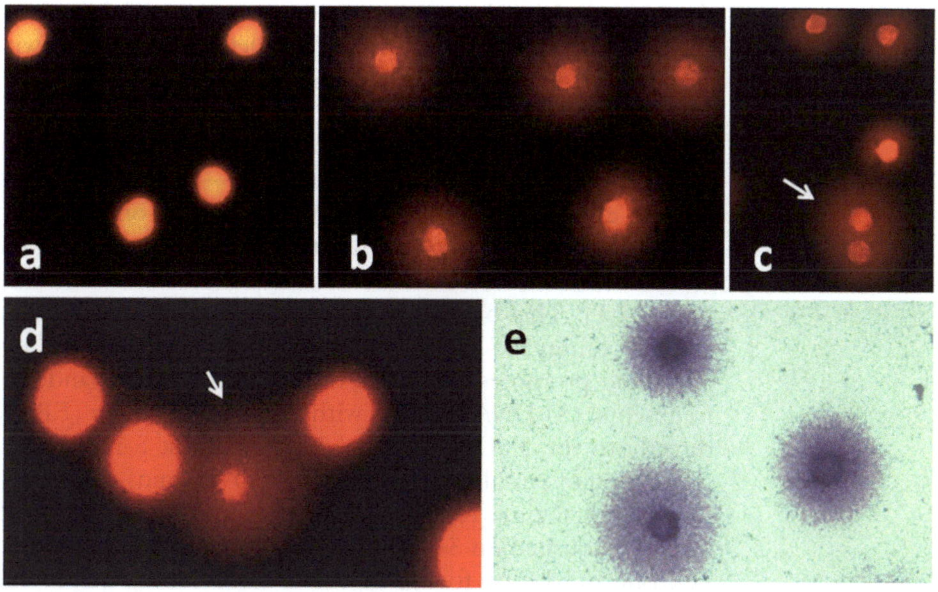

Fig. 1 Representative micrographs of FHA-processed cells. Control (**a**), H_2O_2- (**b**, 100 μM for 20 min) and tert-butylhydroperoxide-treated (**c**, 150 μM for 20 min) human leukemia Jurkat cells; staurosporine- (**d**, 200 nM for 6 h) and H_2O_2-treated (**e**, 50 μM for 20 min) U937 human promonocytic cells. Panels (**a**)–(**d**) show EB stained cells seen at the fluorescence microscope and (**e**) Wright-stained cells at the transmission light microscope. *Arrow* in (**c**) shows "overlapped" halos and that in (**d**) an apoptotic cell—among three undamaged ones—characterized by a small nuclear cage with most of its DNA diffused in a very large halo. Magnification: 200×

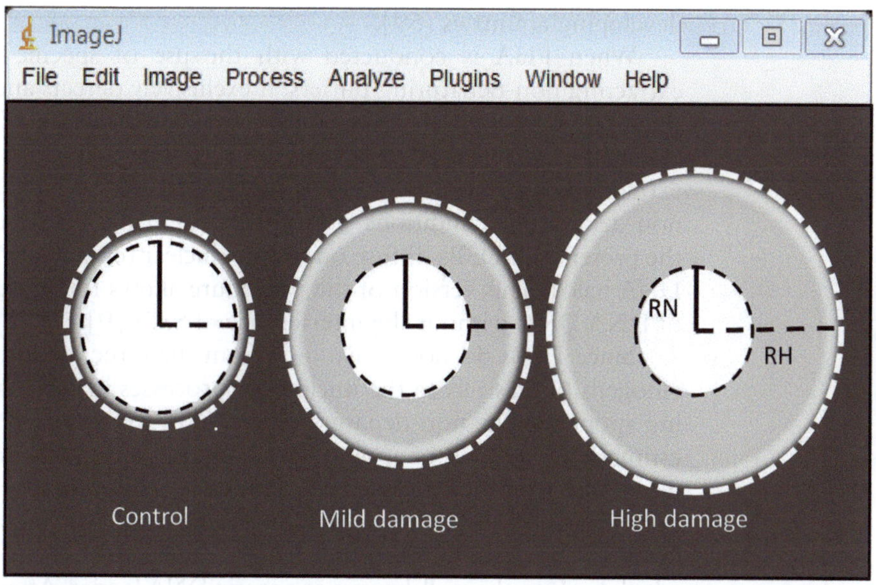

Fig. 2 Representative screenshot of FHA-assayed cells with different degrees of DNA damage analyzed with the ImageJ program. *RN* nucleus radius, *RH* halo radius. Also shown are the circular selections tools used to mark the circumference of the halos and of the nuclei

demonstrated by dose/response experiments and by comparing these results with the outcomes of parallel determinations with the comet assay taken as a reference method [9, 35].

The experimental steps to isolate nuclear DNA and analyze SSBs—including those derived from the conversion of a basic site under alkaline denaturing conditions [11]—in FHA consist of: (1) agarose embedding of cells to form a thin layer on a fully frosted microscope slide; (2) deproteinization of agarose embedded nuclei and denaturation of DNA with strong alkaline solutions (NaOH, pH >12.1). This brief (15 min) incubation in a strong alkaline solution causes the deproteinization of the nuclei, DNA denaturation and promotes the radial diffusion of single stranded DNA fragments; (3) DNA staining with ethidium bromide (EB) added directly to the NaOH solution during the last 5 min of alkaline incubation of samples; (4) destaining (10–15 min).

The total time required to prepare and process up to 9–12 samples mounted on 3–4 fully frosted slides is around 40–45 min, i.e., one third/one fourth of the average time required for the comet assay. Apart from its rapidity, another significant advantage of FHA is that no particular equipment or reagent, except those usually included in a standard tissue culture and cellular biology laboratory, is required. As an example, only NaOH and EB are needed to promote extraction of single stranded DNA fragments and staining. Consequently, the total cost of each experiment is very low. This unique combination of economy and simplicity can make FHA an attractive option for genotoxicity tests by environmental control agencies with limited budgets, such as those in developing countries [59].

When FHA is conducted with the use of specific lysing/extracting non-denaturing buffers, it exclusively detects and quantifies DNA DSBs [9], which are highly cytotoxic [45, 60]. The selectivity of detection is achieved by reducing the pH of the extraction/lysis buffer to a lower value (pH 10.1) [9, 50]. In such non-denaturing conditions only DNA fragments resulting from the presence of DSBs diffuse out of the nuclear remnants and form DNA halos. This version of the procedure allows for quantitation of DNA DSBs without the interference of SSBs [9].

Since DSBs do not derive only from the direct action of carcinogens but also from the nucleolytic processes in cells undergoing apoptosis, the non-denaturing version of FHA can be used to estimate the number of apoptotic cells [9, 50] within a given sample. Under non-denaturing FHA conditions, apoptotic cells produce characteristic, very large-size halos (Fig. 1d). Such highly extended halos are likely to reflect the intense nucleolytic activity which leads to the well-known apoptotic DNA ladder. After counting at least 100 cells per treatment condition, the percent ratio between the number of cells showing large halos and that of intact ones represents an index of the extent of apoptosis. When studying apoptosis, a simple count of "haloed" vs. intact cells is sufficient and no image analysis is needed.

An example of an apoptotic cell is shown in Fig. 1d, where apoptosis was induced with the non DNA-damaging compound staurosporine [61] to avoid the possible interference of frank DNA breaks. This interference may depend on the fact that many apoptogenic agents (e.g., etoposide, anthracyclines, bleomycin) are also capable of inducing direct DNA damage via double-strand breakage, which could be confused with apoptotic DNA cleavage. Even though in most cases these two events do not coincide in time—frank DNA breakage being an early event and apoptotic fragmentation a later one—under particular conditions (i.e., continuous exposure to genotoxins) the simultaneous presence of frank DNA strand breaks and apoptotic DNA cleavage may occur [62]. Hence, under such conditions it might be subjective and difficult to distinguish apoptotic cells from non-apoptotic DNA-damaged ones using only FHA. These potentially ambiguous situations have been specifically discussed elsewhere [9, 63] and can be addressed using FHA with independent techniques capable of distinguishing and quantifying the simultaneous occurrence of more phenomena (e.g., vital stains exclusion and cell growth assays, Hoechst 33,342 staining, caspase colorimetric assay kits, pulsed-field gel electrophoresis).

In the past few years three independent research groups further implemented and significantly improved FHA. Galaz-Leiva et al. [20] developed a new DNA staining procedure that exchanges EB for Wright's methylene blue-eosin (*see* Fig. 1e). This modification allows analysis of DNA with light microscopes instead of fluorescence ones, further simplifies the instrumentation requirements, and reduces the expenses to set up the assay. Moreover, the optical quality of Wright-stained samples is unaffected by the fading and photobleaching typical of EB-stained samples, a factor that helps to perform a more accurate image analysis.

Using a different technique altogether, Quiao et al. [21] developed an array of microfabricated patterns capable of capturing cells through electrostatic interactions. Patterned cells are then trapped inside a layer of agarose gel and assayed according to FHA for DNA damage. The halo array, called HaloChip, can be observed fluorescently after labeling DNA with EB. The regular spatial and focal arrangement of each cell in the array enables automatic estimation of DNA damage. Indeed, one of the major troubles affecting the DNA dispersion assays and FHA is the fact that cells sometimes embed in the agarose film very close to each other or in multiple focal planes thus producing overlapping halos that are difficult to analyze (*see* Fig. 1c). The spatial regularity of HaloChip resolves this problem.

The more recent and probably more significant contribution to improve FHA has been introduced by Maurya et al., who developed HaloJ, a program for the semiautomatic image analysis of

FHA-processed cells [35]. If not assisted by dedicated programs, the analysis of the images from DNA dispersion methods such as comet or FHA is a cumbersome and long procedure per se, requiring even longer times as compared to those needed to complete the assay itself, especially when a high number of samples are prepared. Although the image analysis of FHA, by virtue of the symmetric and regular shape of the halos, is theoretically simpler as compared to comet, where damaged DNA lies within irregularly shaped tails, the availability of specific programs to analyze tails significantly reduces the time needed to complete image processing. On the contrary, the unavailability of corresponding suitable user-friendly programs for FHA has probably limited its popularity despite its economy and rapidity. Indeed, most of the researchers were analyzing halo images manually using generic image analysis freewares (ImageJ). The HaloJ program addresses this problem [35]: it is a semiautomatic program that analyzes halo images in two simple steps and gives the percentage of DNA in halos and nuclei in a total time that can be estimated in one fourth of that needed for manual processing. Importantly, halo parameters (percentage of DNA in halo nuclei and Nuclear Diffusion factor, NDF) obtained with HaloJ matched the comet parameters (percentage of DNA in tail and Olive tail moment) obtained with the Comet Assay Software Project.

HaloJ program significantly speeds up the analysis of halos and is particularly useful for researchers working on the DNA damage and repair, radiation biology, toxicology, cancer biology, and pharmacology.

In conclusion, FHA is the most rapid and inexpensive method to measure DNA damage at the single cell level. Despite its striking simplicity, FHA is as sensitive, flexible, and reliable as other methods for studying DNA lesions and repair.

2 Materials

2.1 Preparation of Ready-to-use Agarose-Coated Slides

1. Regular fully frosted microscope slides (26 × 76 × 1.5 mm).
2. Absolute ethanol.
3. Plastic tweezers with flat tips.
4. Normal melting (NM) agarose. Prepare 100 mL (*see* **Notes 1** and **2**) NM agarose in distilled water (1% w/v): microwave or heat until near boiling and the agarose dissolves. NM agarose left can be stored at 4 °C and reused for up to 2 weeks.
5. Permanent marker or nail paint.
6. Berzelius tall form 100 mL beaker.
7. Slide tray.
8. Aluminum foil.

2.2 Isolation of Cell Samples and Pilot Experiments

1. 100 mM H_2O_2 freshly prepared stock solution in water. For experiments, dilute stock H_2O_2 to the proper concentrations in complete culture medium.
2. 100 mM etoposide stock solution in dimethylsulphoxide (this solution is stable for up to 1 month at room temperature). For experiments, dilute stock etoposide to the proper concentrations in complete culture medium.
3. U937 and HeLa cells (4×10^5 per experimental condition).
4. RPMI 1640 medium with 10% v/v fetal bovine serum and antibiotics (1% standard penicillin-streptomycin solution); to culture U937 cells.
5. MEM medium with 10% fetal bovine serum and antibiotics (1% standard penicillin-streptomycin solution); to culture HeLa cells.
6. Sterile, disposable cell-culture plasticware such as 75 cm^2 flasks for routine growth, 35 mm diameter dishes for experiments, and 10 mL pipettes.
7. CO_2 incubator for cell culture.
8. Phosphate-buffered saline solution containing EDTA (PBS/EDTA): 137 mM NaCl, 1.4 mM Na_2HPO_4, 4.3 mM KH_2PO_4, 2.7 mM KCl, containing 5 mM ethylenediaminetetraacetic acid disodium salt (EDTA), pH 7.4. This solution may be autoclaved and dispensed in 50 mL aliquots.
9. Solution of 0.25% trypsin containing 1 mM EDTA for cells growing as attached monolayers.
10. Bürker hemocytometer.
11. Trypan blue solution (0.4% w/v in distilled water). Prepare 25 mL and filter-sterilize through a 0.22 μm syringe filter. Store refrigerated for up to 3 months.
12. Large plastic tray filled with 3–4 L of flake ice.

2.3 Agarose Embedding of Cell Samples

1. Low Melting (LM) agarose. Prepare 50 mL stock 0.75% LM agarose (375 mg for 50 mL) in PBS. Microwave or heat until near boiling and the LM agarose dissolves. Aliquot LM agarose in 10 mL glass scintillation vials, cap and store refrigerated until needed. Sealed aliquots are stable for at least 6 months at 4 °C. Single aliquots can be melted and frozen many times but, in our experience, their post-opening stability decreases to 3–4 weeks.
2. Eppendorf-type microfuge conical tubes (1.5 mL).
3. Ready-to-use agarose-coated slides.
4. Coverslips ($22 \times 22 \times 0.15$ mm).
5. Aluminum foil.
6. Large plastic tray filled with 3–4 L of flake ice.

2.4 Sample Lysis and Staining (Denaturing Alkaline FHA for the Assessment of DNA SSBs)

1. Lysis-denaturing-extracting solution: 300 mM NaOH, pH ≥ 13.0. Prepare 1 L stock solution and store at room temperature. If tightly capped, NaOH is stable for up to 1 year.
2. 2.5 mg/mL ethidium bromide (EB) in distilled water (100× stock solution) (*see* **Note 3**).
3. Wright's methylene blue-eosin stain solution (Merck).
4. PBS.
5. DePeX (Serva).
6. Three staining jars and one staining rack with horizontal tray.

2.5 Sample Lysis and Staining (Non-Denaturing Alkaline FHA for DNA DSBs)

1. Lysis-extracting buffer: 150 mM NaOH, 100 mM NaH_2PO_4, 1 mM EDTA free acid, Triton X100 1% v/v, pH 10.1. Prepare 250 mL and store at room temperature. Ensure the bottle is tightly capped. This solution is stable for up to 1 month.
2. PBS (50 mL) containing 0.1 mg/mL RNase (Type 1A ribonuclease from bovine pancreas) (PBS/RNase). Prepare 100× RNase stock aliquots (10 mg/mL) in PBS, and store at −20 °C. RNase aliquots are stable for at least 6 months.
3. 2.5 mg/mL ethidium bromide (EB) in distilled water (100× stock solution) (*see* **Note 3**).
4. Wright's methylene blue-eosin stain solution (Merck).
5. PBS.

2.6 Microscopic Examination and Imaging

1. Depending on the stain procedure (EB or Wright stain), fluorescence or transmitted light microscope, with standard C-mount adapter (*see* **Note 4**).
2. Standard C-mount digital camera coupled and connected to a personal computer (*see* **Note 4**).

2.7 Evaluation of DNA Damage

Image analysis software: we recommend ImageJ running HaloJ [35], the dedicated macro developed by Prof. D.K. Maurya to easily process FHA images. Both programs are freely available: Image J can be downloaded from http://rsbweb.nih.gov/ij/ and HaloJ can be requested to the author according to http://ijt.sagepub.com/content/33/5/362.

3 Methods

The choice between the denaturing or the non-denaturing version of FHA depends on the type of DNA lesion or event to be studied. In particular, the non-denaturing FHA is suitable for studying the levels of DNA DSBs without the interference of background DNA single-strand breakage. Indeed, denaturing FHA (Subheading 3.4) detects either DNA-SSBs or DNA-DSBs, while non-denaturing FHA (Subheading 3.5) selectively scores the presence of DNA-DSBs.

3.1 Preparation of Ready-to-Use Agarose-Coated Slides

Ready-to-use agarose-coated slides should be prepared before experiments. As 6–9 samples can be handled in each experiment, and three samples can be accommodated on a single slide, 20 agarose-coated slides will suffice for 6–10 experiments.

1. Wash 20 slides with absolute ethanol and let them dry at room temperature. Label the upper left corner of each slide (the side to be coated) with permanent ink marker or with nail paint.
2. Melt 1% NM agarose in a microwave oven (*see* **Note 2**).
3. Pour 1% NM agarose in the tall form beaker when it is still hot (55–60 °C). Firmly hold each slide using the tweezers, dip it into the beaker, and slowly remove. Repeat this procedure three times per each slide.
4. Wipe underside of slides to remove agarose. Put the slides (coated side up) in a tray on a flat surface and let them dry overnight at room temperature. For quicker drying, slides can be heated for 3 h at 50 °C in a conventional oven, or microwaved for 5–10 min (set the microwave at the minimum power).
5. Wrap the slides with an aluminum foil and store at room temperature until needed. Ready-to-use slides can be stored for up to 3 months. Avoid high humidity conditions that may promote bacterial contamination of the agarose films.

3.2 Isolation of Cell Samples and Pilot Experiments

Cell samples should be collected immediately after treatments, and handled under dim light to prevent UV-induced DNA damage. It is also worth considering that some types of DNA lesions, such as X-ray-induced DNA-SSBs, are rapidly repaired by living cells. Thus, to block unwanted resealing of DNA breaks, it is mandatory to chill cells immediately after treatments. This can be easily achieved by placing plastic dishes or tubes containing cells on ice, by using prechilled solutions and by presetting the refrigerated centrifuge at 4 °C. The procedures described hereafter refer to cultured cells (*see* **Note 5**).

1. To test and tune up FHA, pilot experiments can be performed. These should include standard treatments producing DNA SSBs or DNA DSBs to test denaturing FHA (*see* Subheading 3.4) or non-denaturing FHA (*see* Subheading 3.5), respectively. X-rays represent the most reliable and affordable DNA damaging agent which dose-dependently induce either DNA-SSBs or DSBs in the range 0.1–10 grays. Alternatively, DNA-SSBs can be generated by treatment with increasing concentrations of H_2O_2 (0.05–0.5 mM for 15 min in complete culture medium at 37 °C); DSBs can be generated by exposure to increasing concentrations of etoposide (10–100 μM for 90 min at 37 °C). The concentrations of both agents should result in DNA breakage ranging from low to extensive damage (*see* Fig. 1 and Subheadings 3.7 and 3.8). These experiments

can be performed using any common cultured cell line sensitive to the above agents, such as U937 human promonocytes, as a reference suspension cell line, or HeLa cells, as a reference adherent cell line. Recently, Galaz-Leiva et al. [20] suggested the use of nucleated erythrocytes from chicken blood as reference cells because they are a convenient and easily available model of terminally differentiated nucleated cells with constant DNA content, chromatin structure, and physiology (no transcription, no replication). Once treated, cells will be immediately processed as described hereafter.

2. Suspension cells: place cell suspension in 15 mL plastic centrifuge tubes and harvest cells by centrifugation at 250 × *g* for 5 min.
3. Discard the supernatant and wash cell pellet with 5 mL ice-cold PBS/EDTA. Centrifuge at 250 × *g*, for 5 min, discard the supernatant, and repeat washing with 5 mL PBS/EDTA (*see* **Note 6**).
4. Before the last centrifugation, resuspend the cells, take a 100 μL aliquot of the suspension, and mix with an equal volume of 0.4% trypan blue. Count cells to determine their density and viability with the Bürker hemocytometer. Since the presence of necrotic cells may profoundly affect the experimental outcomes [9], the number of non-viable, trypan blue-stained cells in treated samples should not be significantly different from that in control samples.
5. Adherent cells: place culture dishes on a tray filled with flake ice and harvest cells by trypsinization (*see* **Note 7**). When cells detach quench trypsin with 1/4 volume of complete culture medium.
6. Place trypsinized cells in 15 mL plastic centrifuge tubes. Wash the cells with 5 mL PBS/EDTA. Centrifuge at 250 × *g* for 5 min, pour off supernatant, and repeat washing (*see* **Note 6**). Before the last centrifugation, count cells and determine their number and viability as in Subheading 3.2, **step 4**.
7. After the last centrifugation, discard supernatants and resuspend pellets at 1 × 10^6 cells/mL in ice-cold PBS/EDTA.
8. Store the samples on ice until casting (stick the plastic tubes in flake ice).

3.3 Agarose Embedding and Sample Casting

The samples that have been prepared according to Subheading 3.2 need to be mixed with LM agarose and spread on the agarose-coated slides.

1. Melt an aliquot of LM agarose in microwave (or by another appropriate method) 10–15 min before casting the samples. Place the LM agarose vial in a 40 °C water bath to cool and stabilize the temperature.

2. Put 40 μL of cell suspension (~3 × 10^4 cells) in an Eppendorf tube and add 40 μL melted LM agarose. Thoroughly mix cells with LM agarose by gently pipetting up and down two to three times (a 100–200 μL micropipette set at 50 μL works fine).
3. Immediately pipette a drop (30–40 μL) of the mixture on the agarose coated slide. Rapidly cover with a coverslip.
4. Repeat **steps 2** and **3** for the next two samples (up to three samples can be accommodated on a single agarose-coated slide).
5. After loading the three samples, put the slides on an aluminum foil resting on ice until the agarose layers harden (~10 min).
6. As suggested in [20] the procedure described in Subheading 3.3, **steps 3–5** can be substituted with simply making smears of the cell suspensions obtained as described in step Subheading 3.3, **step 2** on glass slides.

3.4 Lysis and Staining of Samples Under Denaturing Conditions Using EB

Equilibrate all the solutions at room temperature before use.

1. Add 40 mL (or enough to cover the slides) of 300 mM NaOH to the first staining jar.
2. Gently remove the coverslips (sliding and rotating will help); ensure the samples are firmly stuck on it.
3. Place each slide in the horizontal staining tray.
4. Slowly lower the tray into the first jar with NaOH. Incubate slides for 15 min.
5. Add 0.4 mL 100× EB (final concentration 25 μg/mL) to NaOH during the last 5 min of incubation. Gently shake the staining jar to disperse EB.
6. Pass the staining tray to the second jar containing 40 mL (or enough to cover the slides) distilled water and destain for up to 15 min.
7. After destaining, store the slides at room temperature in the third jar with 40 mL (or enough to cover the slides) fresh distilled water to minimize EB contaminations. Up to 2 h storage in distilled water does not affect halo size.

3.5 Lysis and Staining of Samples Under Non-Denaturing Conditions Using EB

This version of FHA allows selective measurement of the level of DNA DSBs. Since the lysis and extraction of samples are conducted under non-denaturing pH conditions, the two DNA strands are not separated, no single-stranded DNA fragment is released and the presence of DNA SSBs is not revealed by the assay [9]. This is particularly useful when testing agents known or suspected to induce DNA DSBs. This version of FHA is also suitable to unravel the presence, in a population of cells treated with cytotoxic, non-genotoxic agents, of necrotic and/or apoptotic cells [9], whose DNA is highly fragmented by the activity of endonucleases which mostly generate DSBs [46].

Equilibrate all the solutions at room temperature before use.

1. Add 40 mL (or enough to cover the slides) of lysis-extracting solution to the first staining jar (*see* Subheading 2.5, **item 1**).
2. Gently remove the coverslips (sliding and rotating will help) and carefully peel off the casting spacer from the slide; ensure the samples are firmly stuck on it.
3. Place each slide in the horizontal staining tray.
4. Slowly lower the tray into the jar with the lysis-extracting solution.
5. After 10 min of lysis, take out the slides and gently wash them with 3 mL flowing PBS per slide to remove excess detergents and alkali. This will prevent RNase inactivation (*see below*).
6. Dilute 0.4 mL 100× RNase (final concentration 0.1 mg/mL) in 40 mL PBS. Pour this solution in the second staining jar.
7. Slowly lower the slides into the jar with PBS/RNase and incubate for further 15 min. This step is required because RNA is not fully digested at pH values ≥10.1 (the persistence of RNA would result in the formation of large halos of undegraded RNA in either control or treated cells, and would obviously affect the experimental outcomes).
8. Add 0.4 mL 100× EB (final concentration 25 μg/mL) to PBS/RNase during the last 5 min of incubation. Gently shake the staining jar to disperse EB.
9. Pass the slides to the third jar containing distilled water and destain for up to 15 min.
10. Once destained, store the slides at room temperature in a jar (rinse thoroughly the first jar and re-use it) containing fresh, distilled water to minimize EB contaminations. Up to 2 h storage of slides in distilled water does not affect halo size.

3.6 Alternative Staining of Halos for Bright Field Microscopy

Staining of halos to be observed under bright-field microscopy can be performed using commercial Wright's methylene blue-eosin stain solution freshly diluted 1:1 with PBS.

1. Following treatment with the extracting buffers (pH 13.0 or 10.1) and RNase (pH 10.1 procedure only), wash the slides with PBS.
2. Dilute 1:1 Wright's methylene blue-eosin stain solution in PBS; filter the resulting aqueous staining solution and apply to the slides for 10 min at room temperature. Wash slides in tap water for 5 min and examine immediately at the microscope or air dry and eventually mount in DePeX to obtain permanent preparations.

3.7 Microscopic Examination and Image Acquisition

1. Take out one slide at a time from distilled water and, before examination, blot away excess liquid on the back and edges; then put a coverslip over the agarose layers containing the cells.
2. For visualization of DNA halos and nuclei, observations of EB- or Wright-stained DNA are made using a 20–25× objective on a fluorescence or bright field microscope, respectively. Depending on the size of cells, other objectives (16×, 40×) can be used.
3. Capture the images of at least 75–100 cells per experimental condition, i.e. 5–10 observation fields with 10–15 cells each (*see* **Notes 8** and **9**).
4. When finished to score the first slide, put it back into distilled water.
5. Repeat these steps for the other slides (*see* **Notes 10** and **11**).

3.8 Evaluation of DNA Damage

Microscopic images (representative micrographs are shown in Fig. 1) are digitally recorded on a PC coupled to a digital camera and processed with an image analysis software. We use the ImageJ freeware (the Windows version of the original NIH Image) running the HaloJ macro [35].

1. Evaluation is done semiautomatically by HaloJ, which measures the area of the nuclear remnant (nucleus) and that of the corresponding halo (total area of nucleus plus halo) of each scored cell. A flow chart of the image analysis with HaloJ is shown in Fig. 3. Although much less accurate, other even quicker methods can be adopted (*see* **Note 12**).
2. The level of DNA breakage in each scored cell is conveniently expressed by the NDF, which is calculated by HaloJ and represents the ratio between the (total area of nucleus plus halo and the area of the nucleus) -1. The subtraction of one unit from the areas ratio has been introduced because ideal control cells with no damage show no halos: in other words the area of the halo coincides with that of the nucleus, and their ratio is 1. However, since an NDF value equal to 1 as referred to undamaged cells would be confusing, we have introduced this normalization procedure (*see* **Note 13**).

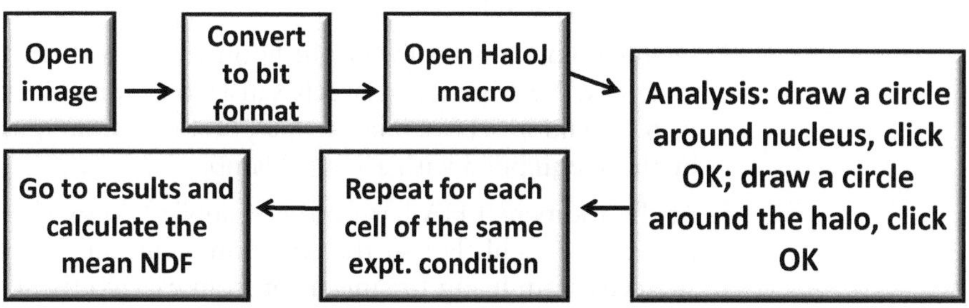

Fig. 3 Flow chart of the sequential operations to analyze FHA cells in HaloJ macro

3. The NDF values of 50–75 randomly selected cells per experimental condition should be calculated. Thus, the NDF value of a given sample, i.e. the level of DNA damage in that experimental condition, is obtained by calculating the mean NDF value of all the scored cells in that sample. As an example, cells treated with the concentrations of H_2O_2 or etoposide indicated in Subheading 3.2, **step 1** should result in NDF values ranging from 1 to 10.
4. Negative (untreated cells) and positive (cells subjected to a standard DNA-damaging treatment) control samples can be included in each experiment (*see* **Notes 13** and **14**).
5. HaloJ automatically exports each cell's NDF to a "Result" sheet, that can be opened and processed with Excel to easily calculate mean NDF values, standard deviations and standard error means in each sample.
6. Statistical analysis is based on multiple experiments from multiple cultures.
7. The NDF values can be conveniently used to generate dose/response or damage induction/repair kinetic plots.

4 Notes

1. The amounts of NM agarose indicated in Subheading 2.1, **item 4** are based on using 26 × 76 mm slides. Proportional volumes can be used for slides of different size.
2. If the NM agarose layers are not sticking to the slides properly, increasing its concentration to 1.5% should eliminate the problem.
3. Due to its intercalating capacity, EB is classed as a mutagen and possible carcinogen and teratogen. The dye should be handled carefully and properly to avoid accidental exposure. Also, EB solutions must not be disposed of without deactivation or decontamination. Deactivation can be obtained by overnight bleaching or by the procedure described in [64]. For decontamination, specific destaining bags and extractors are commercially available.
4. Digital color-cameras for bright fluorescence with a resolution of at least 1.3 Mpixel are largely satisfactory for FHA imaging. As to the fluorescence microscopes, best results are guaranteed by those equipped with a 100 W lamp.
5. At the moment FHA has not been applied to ex vivo tissues. However, it is likely that the common techniques used to isolate cells from living organisms or from ex vivo tissues for the comet assay [65] are suitable for FHA as well.

6. Ensure complete removal of cell culture media before dropping cells on the agarose-coated slide, as it can affect the adhesion of the LM agarose layer to the NM agarose-coated slide.
7. Since excessive trypsinization times may increase DNA damage, it is preferable to harvest adherent cells refractory to this process as follows: gently scrape monolayers in 1 mL PBS/EDTA with a Teflon scraper; disperse cell aggregates by repeatedly pipetting up and down with a 1000 μL micropipette or a Pasteur pipette and then proceed as indicated in Subheading 3.2, **steps 6–8**.
8. Before capturing the images of all the samples of any experiment, adjust and preset the shutter aperture and the exposure time of the digital camera to obtain good images of both control and treated samples.
9. As the majority of the assays based on nonautomated image analysis, FHA data may be affected by operator-bias. To minimize operator-bias, a single inspector blind to the identity of the samples should be allowed to perform the microscopic examination and the analysis.
10. The optical quality of specimens is critical to make good halos and nuclei measurements. Since we noticed that, in the case of EB staining, the best image quality is assured by freshly prepared, wet samples, do not let agarose layers dry before the first microscopic examination has been completed. This can be avoided by storing slides in distilled water, and by observing one slide at a time.
11. As an alternative, the slides can be fixed in alcohol and stored for future observations. After scoring all the slides, remove coverslips, rinse in 100% alcohol, let it dry, and store in dry area. When convenient, flood the dried slides with 500 μL of distilled water, cover with a coverslip, remove excess liquid, and view samples under the microscope. If EB fluorescence is too low, slides can be stained again with an aqueous solution of 25 μg/mL EB for 5–10 min, and destained in distilled water.
12. Although the use of HaloJ is greatly recommended, other approaches similar to those developed by some authors for comet assay [66] can be used. For example, the frequency of cells with no, small, medium, large, and extra-large halos can be counted among 100 cells; scores ranging from 0 to 4 can be attributed to each category and used to quantify DNA damage. These semiquantitative operations, which significantly speed up the analysis procedure, can be done directly at the microscope, without recording the images.
13. The NDF value of untreated cells should be always less than 1.0 (i.e., control cells should display thin halos). Experiments where untreated control cells show higher NDF values are

indicative of badly executed operations and/or badly cultured cells, and should be discarded. In this regard, routinely check cells for mycoplasmal infections: infected cells, although untreated, show large halos that depend on the release of mycoplasmal DNA. Also, discard all solutions and prepare fresh ones and precoat new slides with fresh agarose.

14. To evaluate whether the assay is functioning correctly, a negative (untreated) and a positive (damaged) control should be included in each experiment. As positive controls for DNA SSBs we use cryopreserved U937 cells that were pretreated for 15 min with 150 μM H_2O_2 in PBS at 37 °C. Such a treatment results in extensive DNA single-strand breakage (NDF between 7 and 10) in the majority of cells. This cell population can be stored immediately after oxidative challenge as 0.1–0.3 mL aliquots into cryotubes. To freeze cells accurately wash pellets with fresh, ice-cold PBS/EDTA after H_2O_2 challenge; resuspend at 1×10^6/mL in 10% v/v dimethylsulfoxide, 40% v/v growth medium and 50% v/v fetal calf serum. Transfer aliquots of 0.1–0.3 mL in cryotubes and freeze at −70 °C. For experiments, frozen cells will be recovered by submerging cryotubes in 37 °C water bath until the last trace of ice has melted. Process such cells as indicated in Subheading 3.4 and following. The NDF data resulting from these positive control cells processed along with the studied samples can be used to evaluate, and to take into account, inter-experimental variability.

References

1. Moustacchi E (2000) DNA damage and repair: consequences on dose-responses. Mutat Res 464:35–40
2. Erixon K, Ahnstrom G (1979) Single-strand breaks in DNA during repair of UV-induced damage in normal human and xeroderma pigmentosum cells as determined by alkaline DNA unwinding and hydroxylapatite chromatography: effects of hydroxyurea, 5-fluorodeoxyuridine and 1-beta-D-arabinofuranosylcytosine on the kinetics of repair. Mutat Res 59:257–271
3. Furihata C, Matsushima T (1987) Use of in vivo/in vitro unscheduled DNA synthesis for identification of organ-specific carcinogens. Crit Rev Toxicol 17:245–277
4. Kohn KW, Grimek-Ewig RA (1973) Alkaline elution analysis, a new approach to the study of DNA single-strand interruptions in cells. Cancer Res 33:1849–1853
5. Larsen KH, Brash D, Cleaver JE et al (1982) DNA repair assays as tests for environmental mutagens. A report of the U.S. EPA Gene-Tox program. Mutat Res 98:287–318
6. Ostling O, Johanson KJ (1984) Microelectrophoretic study of radiation-induced DNA damages in individual mammalian cells. Biochem Biophys Res Commun 123:291–298
7. Tice RR, Andrews PW, Singh NP (1990) The single cell gel assay: a sensitive technique for evaluating intercellular differences in DNA damage and repair. Basic Life Sci 53:291–301
8. Sestili P, Cantoni O (1999) Osmotically driven radial diffusion of single-stranded DNA fragments on an agarose bed as a convenient measure of DNA strand scission. Free Radic Biol Med 26:1019–1026
9. Sestili P, Martinelli C, Stocchi V (2006) The fast halo assay: an improved method to quantify genomic DNA strand breakage at the single-cell level. Mutat Res 607:205–214
10. Bacso Z, Eliason JF (2001) Measurement of DNA damage associated with apoptosis by laser scanning cytometry. Cytometry 45:180–186
11. Brigotti M, Alfieri R, Sestili P et al (2002) Damage to nuclear DNA induced by Shiga

toxin 1 and ricin in human endothelial cells. FASEB J 16:365–372

12. Brigotti M, Arfilli V, Carnicelli D et al (2013) Shiga toxin 1, as DNA repair inhibitor, synergistically potentiates the activity of the anticancer drug, mafosfamide, on raji cells. Toxins (Basel) 5:431–444
13. Brigotti M, Carnicelli D, Ravanelli E et al (2007) Molecular damage and induction of proinflammatory cytokines in human endothelial cells exposed to Shiga toxin 1, Shiga toxin 2, and alpha-sarcin. Infect Immun 75:2201–2207
14. Cantoni O, Guidarelli A (2008) Indirect mechanisms of DNA strand scission by peroxynitrite. Methods Enzymol 440:111–120
15. Chaudhary P, Shukla SK, Kumar IP et al (2006) Radioprotective properties of apple polyphenols: an in vitro study. Mol Cell Biochem 288:37–46
16. Chaudhary P, Shukla SK, Sharma RK (2011) REC-2006-a fractionated extract of *Podophyllum hexandrum* protects cellular DNA from radiation-induced damage by reducing the initial damage and enhancing its repair in vivo. Evid Based Complement Alternat Med 2011:473953
17. Crimella C, Cantoni O, Guidarelli A et al (2011) A novel nonsense mutation in the APTX gene associated with delayed DNA single-strand break removal fails to enhance sensitivity to different genotoxic agents. Hum Mutat 32:E2118–E2133
18. Di Pietro A, Baluce B, Visalli G et al (2011) Ex vivo study for the assessment of behavioral factor and gene polymorphisms in individual susceptibility to oxidative DNA damage metals-induced. Int J Hyg Environ Health 214:210–218
19. Fimognari C, Turrini E, Sestili P et al (2014) Antileukemic activity of sulforaphane in primary blasts from patients affected by myelo- and lympho-proliferative disorders and in hypoxic conditions. PLoS One 9:e101991
20. Galaz-Leiva S, Perez-Rodriguez G, Blazquez-Castro A, Stockert JC (2012) A simplified chromatin dispersion (nuclear halo) assay for detecting DNA breakage induced by ionizing radiation and chemical agents. Biotech Histochem 87:208–217
21. Garcia-Alonso FJ, Guidarelli A, Periago MJ (2007) Phenolic-rich juice prevents DNA single-strand breakage and cytotoxicity caused by tert-butylhydroperoxide in U937 cells: the role of iron chelation. J Nutr Biochem 18:457–466
22. Godard T, Deslandes E, Lebailly P et al (1999) Early detection of staurosporine-induced apoptosis by comet and annexin V assays. Histochem Cell Biol 112:155–161
23. Grasso S, Scifo C, Cardile V et al (2003) Adaptive responses to the stress induced by hyperthermia or hydrogen peroxide in human fibroblasts. Exp Biol Med (Maywood) 228:491–498
24. Guidarelli A, Cerioni L, Cantoni O (2007) Inhibition of complex III promotes loss of Ca2+ dependence for mitochondrial superoxide formation and permeability transition evoked by peroxynitrite. J Cell Sci 120:1908–1914
25. Guidarelli A, Cerioni L, Fiorani M, Cantoni O (2009) Differentiation-associated loss of ryanodine receptors: a strategy adopted by monocytes/macrophages to prevent the DNA single-strand breakage induced by peroxynitrite. J Immunol 183:4449–4457
26. Guidarelli A, Cerioni L, Tommasini I et al (2005) Role of Bcl-2 in the arachidonate-mediated survival signaling preventing mitochondrial permeability transition-dependent U937 cell necrosis induced by peroxynitrite. Free Radic Biol Med 39:1638–1649
27. Guidarelli A, Clementi E, De Nadai C et al (2001) TNFalpha enhances the DNA single-strand breakage induced by the short-chain lipid hydroperoxide analogue tert-butylhydroperoxide via ceramide-dependent inhibition of complex III followed by enforced superoxide and hydrogen peroxide formation. Exp Cell Res 270:56–65
28. Guidarelli A, De Sanctis R, Cellini B et al (2001) Intracellular ascorbic acid enhances the DNA single-strand breakage and toxicity induced by peroxynitrite in U937 cells. Biochem J 356:509–513
29. Guidarelli A, Palomba L, Cantoni O (2000) Peroxynitrite-mediated release of arachidonic acid from PC12 cells. Br J Pharmacol 129:1539–1541
30. Guidarelli A, Palomba L, Fiorani M, Cantoni O (2008) Susceptibility of rat astrocytes to DNA strand scission induced by activation of NADPH oxidase and collateral resistance to the effects of peroxynitrite. Free Radic Biol Med 45:521–529
31. Guidarelli A, Sciorati C, Clementi E, Cantoni O (2006) Peroxynitrite mobilizes calcium ions from ryanodine-sensitive stores, a process associated with the mitochondrial accumulation of the cation and the enforced formation of species mediating cleavage of genomic DNA. Free Radic Biol Med 41:154–164
32. Guidarelli A, Sestili P, Fiorani M, Cantoni O (2000) Arachidonic acid induces calcium-dependent mitochondrial formation of species promoting strand scission of genomic DNA. Free Radic Biol Med 28:1619–1627
33. Guidi C, Potenza L, Sestili P et al (2008) Differential effect of creatine on oxidatively-injured

mitochondrial and nuclear DNA. Biochim Biophys Acta 1780:16–26

34. Kadioglu E, Sardas S, Erturk S et al (2009) Determination of DNA damage by alkaline halo and comet assay in patients under sevoflurane anesthesia. Toxicol Ind Health 25:205–212
35. Maurya DK (2014) HaloJ: an ImageJ program for semiautomatic quantification of DNA damage at single-cell level. Int J Toxicol 33:362–366
36. Meintieres S, Nesslany F, Pallardy M, Marzin D (2003) Detection of ghost cells in the standard alkaline comet assay is not a good measure of apoptosis. Environ Mol Mutagen 41:260–269
37. Mondal NK, Bhattacharya P, Ray MR (2011) Assessment of DNA damage by comet assay and fast halo assay in buccal epithelial cells of Indian women chronically exposed to biomass smoke. Int J Hyg Environ Health 214:311–318
38. Palomba L, Guidarelli A, Scovassi AI, Cantoni O (2001) Different effects of tert-butylhydroperoxide-induced peroxynitrite-dependent and -independent DNA single-strand breakage on PC12 cell poly(ADP-ribose) polymerase activity. Eur J Biochem 268:5223–5228
39. Potenza L, Martinelli C, Polidori E et al (2010) Effects of a 300 mT static magnetic field on human umbilical vein endothelial cells. Bioelectromagnetics 31:630–639
40. Qiao Y, An JC, Ma LY (2013) Single cell Array based assay for in vitro Genotoxicity study of nanomaterials. Anal Chem 85:4107–4112
41. Qiao Y, Ma LY (2013) Predicting efficacy of cancer cell killing under hypoxic conditions with single cell DNA damage assay. Anal Chem 85:6953–6957
42. Qiao Y, Ma LY (2013) Quantification of metal ion induced DNA damage with single cell array based assay. Analyst 138:5713–5718
43. Qiao Y, Wang C, Su M, Ma L (2012) Single cell DNA damage/repair assay using HaloChip. Anal Chem 84:1112–1116
44. Sestili P, Alfieri R, Carnicelli D et al (2005) Shiga toxin 1 and ricin inhibit the repair of H2O2-induced DNA single strand breaks in cultured mammalian cells. DNA Repair (Amst) 4:271–277
45. Sestili P, Cantoni O, Cattabeni F, Murray D (1995) Evidence for separate mechanisms of cytotoxicity in mammalian cells treated with hydrogen peroxide in the absence or presence of L-histidine. Biochim Biophys Acta 1268:130–136
46. Sestili P, Cattabeni F, Cantoni O (1996) Direct excision of 50 kb pair DNA fragments from megabase-sized fragments produced during apoptotic cleavage of genomic DNA. FEBS Lett 396:337–342
47. Sestili P, Martinelli C, Ricci D et al (2007) Cytoprotective effect of preparations from various parts of *Punica granatum* L. fruits in oxidatively injured mammalian cells in comparison with their antioxidant capacity in cell free systems. Pharmacol Res 56:18–26
48. Sestili P, Paolillo M, Lenzi M et al (2010) Sulforaphane induces DNA single strand breaks in cultured human cells. Mutat Res 689:65–73
49. Shukla SK, Chaudhary P, Kumar IP et al (2006) Protection from radiation-induced mitochondrial and genomic DNA damage by an extract of *Hippophae rhamnoides*. Environ Mol Mutagen 47:647–656
50. Singh NP (2000) A simple method for accurate estimation of apoptotic cells. Exp Cell Res 256:328–337
51. Trivedi PP, Tripathi DN, Jena GB (2011) Hesperetin protects testicular toxicity of doxorubicin in rat: role of NFkappaB, p38 and caspase-3. Food Chem Toxicol 49:838–847
52. Vivek Kumar PR, Cheriyan VD, Seshadri M (2009) Could a strong alkali deproteinization replace the standard lysis step in alkaline single cell gel electrophoresis (comet) assay (pH>13)? Mutat Res 678:65–70
53. Vorob'eva N, Antonenko AV, Osipov AN (2011) Particularities of blood lymphocyte response to irradiation in vitro in breast cancer patients. Radiats Biol Radioecol 51:451–456
54. Wright WD, Lagroye I, Zhang P et al (2001) Cytometric methods to analyze ionizing-radiation effects. Methods Cell Biol 64:251–268
55. Manvati S, Mangalhara KC, Kalaiarasan P et al (2014) MiR-101 induces senescence and prevents apoptosis in the background of DNA damage in MCF7 cells. PLoS One 9:e111177
56. Smetanina NM, Pustovalova MV, Osipov AN (2013) Modified DNA-halo method for assessment of DNA damage induced by various genotoxic agents. Radiats Biol Radioecol 53:389–393
57. Smetanina NM, Pustovalova MV, Osipov AN (2014) Effect of dimethyl sulfoxide on the extent of DNA single-strand breaks and alkali-labile sites induced by 365 nm UV-radiation in human blood lymphocyte nucleoids. Radiats Biol Radioecol 54:169–173
58. Sultan A, Nesslany F, Violet M et al (2011) Nuclear tau, a key player in neuronal DNA protection. J Biol Chem 286:4566–4575
59. Grover IS, Kaur S (1999) Genotoxicity of wastewater samples from sewage and industrial

effluent detected by the allium root anaphase aberration and micronucleus assays. Mutat Res 426:183–188

60. Cantoni O, Sestili P, Cattabeni F et al (1990) Comparative effects of doxorubicin and 4′-epi-doxorubicin on nucleic acid metabolism and cytotoxicity in a human tumor cell line. Cancer Chemother Pharmacol 27:47–51
61. Burattini S, Ferri P, Battistelli M et al (2009) Apoptotic DNA fragmentation can be revealed in situ: an ultrastructural approach. Microsc Res Tech 72:913–923
62. Gichner T, Mukherjee A, Wagner ED, Plewa MJ (2005) Evaluation of the nuclear DNA diffusion assay to detect apoptosis and necrosis. Mutat Res 586:38–46
63. Singh NP (2005) Apoptosis assessment by the DNA diffusion assay. Methods Mol Med 111:55–67
64. Lunn G, Sansone EB (1987) Ethidium bromide: destruction and decontamination of solutions. Anal Biochem 162:453–458
65. Burlinson B et al (2007) Fourth International Workgroup on Genotoxicity testing: results of the in vivo Comet assay workgroup. Mutat Res 627:31–35
66. Ross GM, McMillan TJ, Wilcox P, Collins AR (1995) The single cell microgel electrophoresis assay (comet assay): technical aspects and applications. Report on the 5th LH Gray Trust workshop, Institute of Cancer Research, 1994. Mutat Res 337:57–60

Chapter 7

Rapid Detection of Bacterial Susceptibility or Resistance to Quinolones

Fátima Otero, Rebeca Santiso, María Tamayo, Germán Bou, Jaime Gosálvez, and José Luis Fernández

Abstract

The emergence of multidrug resistant microorganisms together with the decline in discovery and development of new antibiotics is of great concern in health-care policy. In this alarming context, an early and well-tailored antibiotic therapy is a relevant strategy not only to improve clinical outcome but also to avoid appearance and spreading of perilous resistant strains. One of the most common antibiotic classes is fluoroquinolones. They trap the DNA girase and/or topoisomerase IV on the DNA, resulting in DNA fragmentation. We have developed the Micromax® assay to determine, in situ, the integrity of the chromosomal DNA-nucleoid from microorganisms. This was validated as a simple procedure for the rapid assessment of the susceptibility or resistance to quinolones in gram-negative bacteria. After incubating with the quinolone, cells are trapped in an agarose microgel on a slide and incubated with a specific lysing solution to remove the cell wall and visualize the nucleoids under fluorescence microscopy. If the strain is susceptible to the quinolone, the bacterial nucleoids show a halo of diffusing DNA spots of fragmented DNA, whereas they appear intact in the resistant strain. The technical processing is performed in 40 min with practically total sensitivity and specificity.

Key words Bacteria, Antibiotic resistance, Quinolone, DNA fragmentation, Rapid assay

1 Introduction

The emergence and spreading of drug-resistant pathogen strains is considered one of the greatest threats to human health. The European Center for Disease Control (ECDC) reported 25,000 annual deaths due to multi-resistant pathogens. This problem is enhanced by the actual decline in discovery and development of new antibiotics [1]. Microorganisms resistant to multiple antibiotics present a continually increasing health risk, particularly in clinical settings. Specifically, immunocompromised patients and patients located in Intensive Care Units (ICUs) are at increased risk of acquiring nosocomial infectious diseases which may be resistant to one or more antibiotics. In this case, patients may acquire

Vladimir V. Didenko (ed.), *Fast Detection of DNA Damage: Methods and Protocols*, Methods in Molecular Biology, vol. 1644, DOI 10.1007/978-1-4939-7187-9_7, © Springer Science+Business Media LLC 2017

infections through intrusive, but necessary, medical means, such as infections in the respiratory pathway during mechanical ventilation, in the urinary tract or blood vessels via catheters or even through skin wounds, such as incisions required for any number of medical procedures. Such infections may be associated with a high mortality rate. The most habitual pathogens are gram-negative bacilli, frequently multi-resistant to several antibiotics, as *Acinetobacter baumannii*, *Klebsiella pneumoniae*, *Pseudomonas aeruginosa*, and some enterobacteria like *Escherichia coli* [2].

Well-selected, early antibiotic treatment provides the best defense against such multi-resistant pathogens. Current procedures require a bacterial culture for the identification of the microorganism followed by an antibiogram, which routinely requires 2–3 days of bacterial growth. The step of culturing bacteria to construct an antibiogram alone generally requires about 1 day of incubation and is habitually based on the determination of the minimum inhibitory concentration (MIC) of cell growth by the antibiotic [3].

Given the relatively long time necessary to perform the standard antibiogram, antibiotics are usually empirically provided at the onset in life-threatening nosocomial infections. This first line of defense often relies on antibiotics generally known to be effective based on the likely pathogen involved. However, such treatments may be ineffective in 20–40% of cases, and a change of antibiotics later may have a reduced probability of success. Moreover, antibiotic misuse or overuse may result in increasingly resistant strains of bacteria. A fast antibiogram is a big challenge which may save lives and reduce health care costs. Moreover, it may allow us to prevent the spreading of dangerous resistant strains as well as to preserve some strategic antibiotics which sometimes are the only therapeutic option [4–7].

Some of the most extensively prescribed antibiotics are the fluoroquinolones, like ciprofloxacin, norfloxacin, or moxifloxacin. These antibiotics bind to the DNA topoisomerases attached to DNA, DNA gyrase, and/or topoisomerase IV, DNA gyrase being the preferential target in gram-negative bacteria. This binding results in ternary complexes of quinolone-topoisomerase-DNA. After the formation of the ternary complex, a DNA double-strand break (DSB) is produced by the topoisomerase, but bound quinolone inhibits the subsequent ligation of the DNA ends by trapping the topoisomerase. These cleaved complexes are distributed throughout the bacterial chromosome, resulting in DNA fragmentation and subsequent cell killing [8, 9].

Resistance to fluoroquinolones is increasing worldwide, being of great concern in health-care policy. The most habitual mechanism of resistance is the target modification by mutations in the genes encoding the subunits of DNA gyrase or/and topoisomerase IV. Moreover, the intracellular concentration of the drug may be reduced due to changes in drug permeation or overexpression of efflux pumps. A progressively more important mechanism of resistance is the target protection through plasmid-mediated quinolone-resistance genes [10–13]. These different mechanisms may combine resulting in high-level resistance. In any case, resistance to fluoroquinolones results in absence or decrease of the amount of lethal DSBs.

We developed a patented procedure to determine, in situ, the presence of chromosomal DNA fragmentation in microorganisms, at the single-cell level [14–17]. The cells are immersed in an inert microgel on a slide and incubated in a specific lysing solution that removes the cell wall, membranes, and proteins. The nucleoids are stained with a highly sensitive fluorochrome and visualized under a fluorescence microscope. If the bacterial cell is susceptible to the fluoroquinolone, after incubation with the antibiotic it will show the chromosomal DNA fragmented into pieces. The fragments spread from the residual core, showing halos of peripheral diffusion of DNA fragments. The greater the fragmentation, the greater the number of DNA spots and the greater the circular surface area of diffusion. Otherwise, if the bacterium is resistant, the nucleoids appear intact or with much fewer DNA fragments. As of now, the procedure has been successfully validated in 923 clinical isolates from *Escherichia coli* ($n = 163$), *Acinetobacter baumannii* ($n = 322$), *Pseudomonas aeruginosa* ($n = 229$), *Klebsiella pneumoniae* ($n = 189$), *Klebsiella oxytoca* ($n = 4$), *Enterobacter cloacae* ($n = 6$), *Proteus mirabilis* ($n = 5$), and *Proteus vulgaris* ($n = 5$) [16, 17]. The Micromax® assay showed 99.83% sensitivity and 100% specificity in determining resistance to ciprofloxacin, levofloxacin, and norfloxacin.

This technical processing is performed in 40 min. The routine determination of the susceptibility or resistance of gram negative bacteria to fluoroquinolones is established in 80 min, including the 40 min of antibiotic incubation.

2 Materials

2.1 Reagents and Technical Equipment

1. 37 °C incubator (with shaking).
2. 90–100 °C and 37 °C incubation bath(s).
3. 4 °C fridge.
4. Microwave oven.
5. Fume hood.
6. Epifluorescence microscope with appropriate filter for SYBR Gold and objective 100×.
7. Plastic gloves.
8. Micropipettes and tips.
9. Lancet.

10. Plastic tubes (10 or 15 mL).
11. 96-well sterile microtiter plate with lid or 0.5 mL Eppendorf tubes.
12. Trays for horizontal incubations.
13. Glass slide covers (18 × 18 mm or 22 × 22 mm).
14. Culture media: Mueller-Hinton (MH) broth and MH agar plates.
15. Distilled water.
16. Ethanol 70, 90, and 100%.
17. Fluoroquinolone.
18. Micromax kit® prototype (ABM technologies LLC, Navasota, TX). The kit contains coated slides, Eppendorf tubes with low-melting point agarose and a bottle with 125 mL of lysing solution.

2.2 Staining for Fluorescence Microscopy

1. SYBR Gold nucleic acid gel stain (Life Technologies, Carlsbad, CA, USA), 10,000× concentrate in DMSO.
2. Tris Borate EDTA (TBE) buffer 1× (0.09 M Tris-borate, 0.002 M EDTA, pH 7.5).
3. Staining solution: SYBR Gold (2.5 μL/mL) in TBE buffer.

3 Methods

3.1 Bacterial Culture and Incubation with the Antibiotic

1. Select a colony from a 16–24 h plate culture and grow in 2 mL of Mueller-Hinton broth for 2 h at 37 °C with shaking (*see* **Note 1**).
2. Dilute the bacterial culture in culture medium to a concentration of 1 × 10^8 cells/mL. It can be done using a spectrophotometer or a nephelometer (*see* **Note 2**). When a spectrophotometer is used, 1 × 10^8 cells/mL are equivalent to an optical density at 600 nm (OD_{600}) of 0.1. When a nephelometer is used, 1 × 10^8 cells/mL are equivalent to 0.5 McFarland units.
3. Use a sterile 96-well microtiter plate with lid or 0.5 mL Eppendorf tubes for the incubation of bacteria with the antibiotic (*see* **Note 3**). For each strain to be analyzed, two wells (or two tubes) are required: one for the control without antibiotic and another to incubate bacteria with the dose of antibiotic corresponding to the breakpoint of susceptibility, according to the indications of the criterion of the Clinical and Laboratory Standards Institute (CLSI) or the European Union Committee on Antimicrobial Susceptibility Testing (EUCAST).

To prepare each well combine 24 μL of bacterial culture and 6 μL of an antibiotic solution at the appropriate concentration (or 6 μL of distilled H_2O in the control without antibiotic). Incubate bacteria with the antibiotic for 40 min at 37 °C.

3.2 Micromax® Assay Procedure

1. For the microgel embedding of a cell suspension (*see* **Note 4**), first melt agarose in a low-melting point agarose Eppendorf tube, supplied with the Micromax kit, by putting it in a 90 °C–100 °C water bath. The tube should be placed in a foam float so that the top of the tube is at the level of water. Then, leave the floating tube in a water bath at 90–100 °C, until the agarose melts, i.e., around 5 min. Alternatively, melt the agarose in a microwave oven.
2. Transfer the agarose Eppendorf tube, with the float, to a water bath at 37 °C and incubate for 5 min until the temperature has stabilized (*see* **Note 5**).
3. Add 30 μL of the bacterial culture to the agarose Eppendorf tube and gently mix it with the micropipette, avoiding production of air bubbles.
4. Deposit the cell suspension from the agarose Eppendorf tube onto the coated side of a slide provided in the kit, and cover it with a glass coverslip, avoiding trapping air bubbles. A sample size of 10 or 20 μL is recommended for an 18 × 18 mm or 22 × 22 mm coverslip respectively (*see* **Note 6**).
5. Place the slide horizontally on a cold surface, for example a metal or glass plate precooled at 4 °C.
6. Place the cold plate with the slide in the fridge at 4 °C for 5 min, to allow the agarose to solidify.
7. While the slide is in the fridge, deposit 10 mL of the lysing solution provided in with the kit into a tray with dimensions slightly larger than that of a conventional glass slide (*see* **Note** 7).
8. Remove the coverglass by sliding it gently, and immediately immerse the slide in a horizontal position into the lysing solution. Incubate it there for 5 min at room temperature (22 °C) (*see* **Note 8**).
9. Pick up the slide with the help of a lancet and horizontally immerse it for 3 min in a tray containing abundant distilled water to wash off the lysing solution. Wear gloves.
10. To fix the cells, immerse the slide in the horizontal position in trays with increasing concentrations of cold ethanol (70, 90, and 100%). Incubate for 2 min at each concentration.
11. Dry the slide horizontally at room temperature or in an oven at 37 °C.
12. After drying, place the slide in a microwave oven at 750 W for 4 min. When the slide has cooled, it can be immediately stained

or alternatively may be stored in archive boxes at room temperature in the dark for several months.

13. To stain the slide add to it 100 μL of staining solution (SYBR Gold in TBE buffer, 2.5 μL/mL). Cover the slide with a 24 × 60 mm coverslip and incubate for 2 min in the dark (*see* **Notes 9** and **10**).
14. After the incubation, carefully remove the coverslip and briefly wash the slide in TBE buffer. Then cover with a new 24 × 60 mm coverslip (*see* **Note 11**).

3.3 Analysis and Classification of the Sample

1. Examine the sample using a 100× immersion oil objective. Depending on the level of damage, the nucleoids can be divided into three categories:
   - *Nucleoids without damage*: those that spread loops of intact DNA from the central core (*see* Fig. 1a, c, e, f).
   - *Nucleoids with low level of damage*: those that release halos of loops of intact DNA with few peripheral DNA fragments (*see* Fig. 1d).
   - *Nucleoids with high level of damage*: those that present big halos of spots of broken DNA diffused from the residual core (*see* Fig. 1b).
2. After incubation with the antibiotic dose corresponding to the breakpoint of susceptibility (*see* **Note 12**), the strains are categorized as:
   - *Susceptible strains* (*see* Fig. 1a–d): show nucleoids with fragmented DNA (*see* Fig. 1b, d). Strains with MIC close to the breakpoint of susceptibility show nucleoids with low level of DNA damage (*see* Fig. 1d). The lower the MIC of the strain, the higher the degree of damage of its nucleoids.
   - *Non-susceptible (resistant) strains* (*see* Fig. 1e, f): show intact nucleoids (*see* Fig. 1f).

4 Notes

1. We observed that, depending on whether the culture is in logarithmic or stationary phase, bacteria respond differently to the antibiotic. When bacteria are in logarithmic phase, the response to an antibiotic is more uniform, and is in general higher, than when it is in stationary phase. To eliminate variability and improve the results, we always start from a culture in logarithmic growth phase.
2. Cell density within the agarose matrix should not be excessively high in order to avoid the overlapping of cells, and not too broadly spread to facilitate rapid visualization.

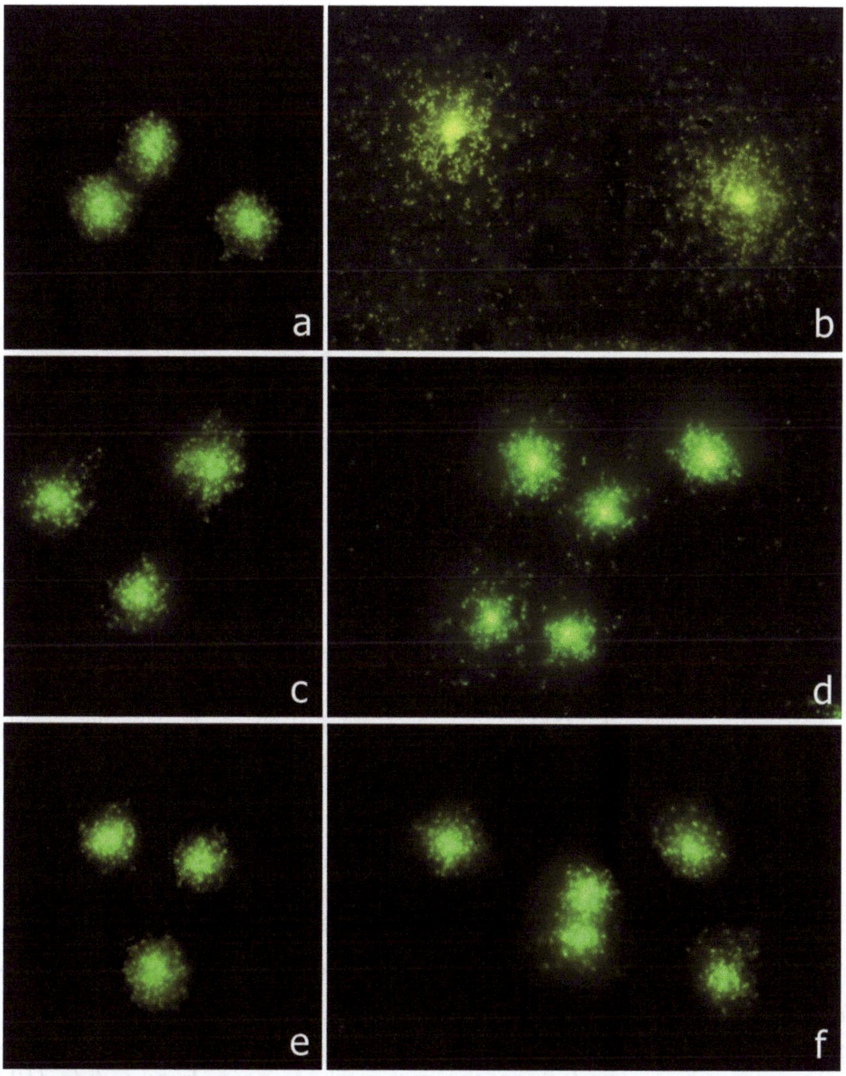

Fig. 1 Three strains of *Escherichia coli* processed to visualize the nucleoids to determine the susceptibility or resistance to ciprofloxacin. The breakpoint concentration of susceptibility for ciprofloxacin in *E. coli*, according to the CLSI criterion [18], is 1 μg/mL. The bacteria were exposed to 0 μg/mL (first column, *left*: **a**, **c**, **e**) and 1 μg/mL of ciprofloxacin (second column, *right*: **b**, **d**, **f**). First strain (first line: **a**, **b**) is susceptible to the quinolone (MIC: 0.012 μg/mL). Second strain (second line: **c**, **d**) is in the limit of susceptibility (MIC: 1 μg/mL). Third strain (third line: **e**, **f**) is resistant (MIC: 4 μg/mL). After incubation with the quinolone dose, the highly susceptible strain shows massively fragmented DNA (**b**), the less susceptible strain reveals a low level of DNA fragmentation (**d**), whereas the nucleoids appear intact in the resistant strain (**f**)

3. The final concentration of antibiotic to be added to the well should correspond to the breakpoint of susceptibility, i.e., the antibiotic concentration used to define isolates as susceptible. The Clinical and Laboratory Standards Institute (CLSI) and the European Union Committee on Antimicrobial Susceptibility Testing (EUCAST) are the international standard-setting

groups who set breakpoint for antibiotics, and they regularly publish guidelines with updated data.

4. The purpose of microgel embedding is to provide an inert support for cells, so they can be processed in a suspension-like environment, but on a slide. This way, they can be easily handled to be incubated in the solutions, avoiding centrifugations. Moreover, possible DNA fragments of relative high size that would be removed to the medium if cells were lysed in suspension, would be retained in the agarose matrix.
5. It is important to mix the cell suspension with the liquid agarose when the latter has stabilized at 37 °C, to avoid cell damage by heat.
6. We recommend using the same slide for both the microgel from the control culture without antibiotic and the microgel from the culture incubated with the antibiotic. This way, both samples will be processed simultaneously, eliminating the possible variability resulting from incubation on different slides.
7. It is recommended to manage the lysing solution under a fume hood, since it is volatile and emits toxic fumes.
8. The lysing step removes the bacterial cell wall and proteins, so the DNA loops that are packed inside the cell spread, producing halos emerging from a central residual core in those bacteria without fragmented DNA. If the DNA is massively broken, the lysing step allows the DNA fragments to diffuse from the residual core producing a big halo of DNA spots. The higher the degree of DNA fragmentation, the bigger the spots halo diffusing from the bacterial core.
9. After being prepared, the staining solution can be stored at 4 °C for a long time.
10. Given the relatively small genome size of microorganisms, classical fluorochromes such as propidium iodide (PI), DAPI, Hoechst 33,258, etc. are not appropriate for staining. In order to visualize the small DNA fragments, it is necessary to use a highly sensitive fluorochrome, such as one from the SYBR family. SYBR Gold provides excellent sensitivity and photostability in comparison with other fluorochromes from the same family, giving an accurate visual assessment under the fluorescence microscope. Antifading solution was not used, since it diminishes the contrast between the small DNA dots and the background.
11. The analysis of the sample under fluorescence microscopy should be done immediately after staining to avoid drying. Alternatively, after staining, the slide may be stored at 4 °C for hours in a humid chamber, in the dark, to prevent drying. If it

dried, the coverslip may be removed by incubation in TBE buffer and, after a brief wash, may be restained again.

12. As an internal control, it is recommended to process a susceptible and a resistant control strain. After processing, nucleoids of the susceptible control strain should reveal its usual DNA fragmentation, whereas nucleoids of the resistant control strain should be intact, thus validating the technique and the results of the other samples.

Acknowledgments

This work was supported by Fondo de Investigaciones Sanitarias (FIS PI14/01346).

References

1. World Health Organization (2014) Antimicrobial resistance: global report on surveillance 2014. http://www.who.int/drugresistance/documents/surveillancereport/en/. Accessed 25 Mar 2015.
2. Peleg AY, Hooper DC (2010) Hospital-acquired infections due to gram negative bacteria. N Engl J Med 362:1804–1813
3. Jorgensen JH, Ferraro MJ (2009) Antimicrobial susceptibility testing: a review of general principles and contemporary practices. Clin Infect Dis 49:1749–1755
4. Iregui M, Ward S, Sherman G, Fraser VJ, Kollef MH (2002) Clinical importance of delays in the initiation of appropriate antibiotic treatment for ventilator-associated pneumonia. Chest 122:262–268
5. Kollef KE, Schramm GE, Wills AR, Reichley RM, Micek ST, Kollef MH (2008) Predictors of 30-day mortality and hospital costs in patients with ventilator-associated pneumonia attributed to potentially antibiotic resistant gram-negative bacteria. Chest 134:281–287
6. Falagas ME, Kasiakou SK, Rafailidis PI, Zouglakis G, Morfou P (2006) Comparison of mortality of patients with Acinetobacter Baumannii bacteriemia receiving appropriate and inappropriate empirical therapy. J Antimicrob Chemother 57:1251–1254
7. Falagas ME, Bliziotis IA (2007) Pandrug-resistant gram-negative bacteria: the dawn of the post-antibiotic era? Int J Antimicrob Agents 29:630–636
8. Drlica K, Malik M, Kerns RJ, Zhao X (2008) Quinolone-mediated bacterial death. Antimicrob Agents Chemother 52:385–392
9. Malik M, Zhao X, Drlica K (2006) Lethal fragmentation of bacterial chromosomes mediated by DNA gyrase and quinolones. Mol Microbiol 61:810–825
10. Chen F-J, Lo H-J (2003) Molecular mechanisms of fluoroquinolone resistance. J Microbiol Immunol Infect 36:1–9
11. Martínez-Martínez L, Pascual A, Jacoby GA (1998) Quinolone resistance from a transferable plasmid. Lancet 351:797–799
12. Robicsek A, Strahilevitz J, Jacoby GA, Macielag M, Abbanat D, Park CH, Bush K, Hooper DC (2006) Fluoroquinolone-modifying enzyme: a new adaptation of a common aminoglycoside acetyltransferase. Nat Med 12:83–88
13. Ruiz J (2003) Mechanisms of resistance to quinolones: target alterations, decreased accumulation and DNA gyrase protection. J Antimicrob Chemother 51:1109–1117
14. Fernández JL, Cartelle M, Muriel L, Santiso R, Tamayo M, Goyanes V, Gosálvez J, Bou G (2008) DNA fragmentation in microorganisms assessed in situ. Appl Environ Microbiol 74:5925–5933
15. Tamayo M, Santiso R, Gosalvez J, Bou G, Fernández JL (2009) Rapid assessment of the effect of ciprofloxacin on chromosomal DNA from *Escherichia coli* using an in situ DNA fragmentation assay. BMC Microbiol 13:9–69
16. Santiso R, Tamayo M, Fernández JL, del Carmen Fernández M, Molina F, Villanueva R, Gosálvez J, Bou G (2009) Rapid and simple determination of ciprofloxacin resistance in clinical strains of *Escherichia coli*. J Clin Microbiol 47:2593–2595

17. Bou G, Otero FM, Santiso R, Tamayo M, Fernández Mdel C, Tomás M, Gosálvez J, Fernández JL (2012) Fast assessment of resistance to carbapenems and ciprofloxacin of clinical strains of Acinetobacter Baumannii. J Clin Microbiol 50:3609–3613
18. Clinical and Laboratory Standards Institute (2014) Performance standards for antimicrobial susceptibility testing; twenty-fourth informational supplement M100-S24, vol 34, 1st edn. Clinical and Laboratory Standards Institute, Wayne, PA

Chapter 8

Rapid Detection of Apoptosis in Cultured Mammalian Cells

Igor Kudryavtsev, Maria Serebryakova, Liudmila Solovjeva, Maria Svetlova, and Denis Firsanov

Abstract

Flow cytometry is a powerful tool for the analysis of apoptosis, the process that directly determines cell fate after the action of different stresses. Here, we describe a flow cytometry method for the assessment of early and late stages of apoptosis in non-fixed cultured cells using SYTO16, DRAQ7, and PO-PRO1 dyes simultaneously. This multicolor flow cytometry procedure requires 45 min for completion and provides a quantitative assessment of cell viability. It can be useful in evaluating the cytotoxic properties of new drugs, and antitumor interventions.

Key words Flow cytometry, Apoptosis, Fluorescent nucleic acid dyes, PO-PRO-1, DRAQ7, SYTO16

1 Introduction

Flow cytometry is one of the best methods to identify and quantify apoptotic cells as well as to study mechanisms associated with this type of cell death. Multicolor flow cytometry allows simultaneous identification of cells at different stages of apoptosis in a single sample tube. The number of fluorochromes having nonoverlapping spectra dramatically increased during the last decade, and it became possible to apply new and easy-to-use multiparameter approaches for apoptosis detection. Multicolor flow cytometry is a low-cost and effective method for evaluation of cytotoxic properties of novel drugs, particularly for antitumor therapy. Quantitative cell viability assessment is necessary for all fields of experimental research in cell biology.

Using three sources of excitation (lasers—with emission wavelengths 405, 488, and 638 nm), three DNA-binding dyes and polychromatic flow cytometry, we have developed a protocol that can be applied to cells undergoing apoptosis. Sample preparation takes less than 45 min. The protocol is based on the application of three DNA-binding dyes—PO-PRO-1, SYTO 16 green, and DRAQ7.

Vladimir V. Didenko (ed.), *Fast Detection of DNA Damage: Methods and Protocols*, Methods in Molecular Biology, vol. 1644, DOI 10.1007/978-1-4939-7187-9_8, © Springer Science+Business Media LLC 2017

YO-PRO-1 is a carbocyanine nucleic acid dye, which penetrates cells through purinergic receptors encoded by P2RX7 gene, which function as ligand-gated ion channels [1]. In living cells, YO-PRO-1 is not accumulated due to almost inactive membrane channels and a very low transporting ability of the membrane. However during the early stages of apoptosis, the channels become activated and the plasma membrane loses asymmetry. These changes lead to the intracellular accumulation of YO-PRO-1. YO-PRO-1 excitation/emission wavelength maxima are 491/509 nm. PO-PRO-1, which is used in this protocol, is similar to the green fluorescent dye YO-PRO-1in its ability to penetrate cells at the early stages of apoptosis, and differs from YO-PRO-1 only in excitation/emission wavelength maxima that are 435/455 nm.

SYTO 16 is a cell-permeant green-fluorescent dye that enhances its fluorescence upon binding nucleic acids. SYTO 16 can stain not only nuclear DNA, but also mitochondrial DNA and cytoplasmic RNA. Viable cells always show intensive SYTO 16 staining, early apoptotic cells are characterized by less intensive staining, and late apoptotic cells lose the ability to bind the dye. The mechanisms of differential staining of viable and apoptotic cells are unclear [2, 3]. Excitation and emission spectra of this dye are similar to fluorescein isothiocyanate (FITC). SYTO 16 has excitation wavelength maxima 488 nm in the presence of DNA, and 494 nm in the presence of RNA, and emission maxima 518/525 nm (DNA/RNA).

Deep red anthraquinone 7 (DRAQ7) is a fluorescent DNA-intercalating dye that does not permeate viable cells. It is able to penetrate cells and nuclei only via damaged/fragmented membranes [4]. DRAQ7 can be used for DNA staining in fixed cells or for discrimination between viable and apoptotic cells. DRAQ7 has excitation wavelength maxima 599/644 nm, and emission wavelength maxima 678/694 nm when the dye is intercalated into double-stranded DNA.

The presented protocol was tested on human THP-1 monocytic leukemia cells treated with camptothecin.

2 Materials

1. THP-1cells (*see* **Notes 1** and **2**).
2. PBS supplemented with 2% fetal bovine serum (FBS)—“wash buffer.”
3. PO-PRO-1 iodide—1 mM solution in DMSO (Thermo Fisher Scientific). Dilute PO-PRO-1 iodide from 1 mM stock solution in DMSO to 2.5 μM in sterile PBS immediately prior to use (*see* **Note 3**).
4. SYTO 16 green fluorescent nucleic acid stain—1 mM solution in DMSO (Thermo Fisher Scientific). Dilute SYTO 16 green

from 1 mM stock solution in DMSO to 2.5 μM in sterile PBS immediately prior to use.

5. DRAQ7 Far-Red Fluorescent Live-Cell Impermeant DNA Dye (Beckman Coulter). Dilute DRAQ7 from 300 μM stock solution to 30 μM in sterile PBS immediately prior to use. Add 10 μL of 30 μM DRAQ7 per 90 μL cell sample to the final concentration of 3 μM.
6. 12 × 75 mm tubes suitable for flow cytometer (in the current example: 5 mL tubes).
7. Flow cytometer with 405, 488, and 638 nm excitation wavelengths and filters for the collection of blue, green, and red fluorescence (in the current example—Navios flow cytometer equipped with violet, blue, and red solid state diode lasers and filters 450 ± 50 nm, 525 ± 40 nm, and 725 ± 20 nm (Beckman Coulter).

3 Method

The duration of the procedure is 45 min.

1. Collect—a sample of untreated and camptothecin-treated cells from a 96-well plate (from each well, take 200 μL of cell suspension in complete culture medium, 1–2 × 10^6 cells/mL) and place them into 12 × 75 mm tubes suitable for flow cytometer.
2. Add 4.5 mL of cold PBS to each tube.
3. Centrifuge cells for 5 min at 300–330 × *g* at room temperature (RT).
4. Gently remove the supernatant by aspiration.
5. Resuspend cell pellet in 100 μL of fresh PBS.
6. Add 10 μL of 2.5 μM of SYTO 16 nucleic acid stain in PBS (*see* **Notes 4** and **5**).
7. Gently vortex the cells and incubate for 15 min at RT in the dark.
8. Add 4.5 mL of cold wash buffer into each tube (*see* **Notes 6** and 7).
9. Centrifuge cells for 5 min at 300–330 × *g* at RT.
10. Gently remove the supernatant by aspiration.
11. Resuspend cell pellet in 100 μL of fresh PBS.
12. Add 10 μL of PO-PRO-1 iodide solution and 10 μL of DRAQ7 solution (these two solutions can be premixed before the addition to the sample).
13. Gently vortex the cells and incubate for 5 min at RT in the dark.
14. Add 300 μL of fresh PBS to make the final volume of 400 μL.

15. Analyze all samples by a flow cytometer.
16. Collect at least 15,000 cells per each sample.
17. Data analysis:
    (a) Using dot plot "forward scatter versus side scatter" ("FS versus SS"), adjust the voltages to be able to see the whole cell population.
    (b) Using samples of cells separately stained with each fluorescent dye, adjust the voltages for fluorescent channels. Using PO-PRO-1-stained sample of cells with induced apoptosis (positive control), adjust the voltage for the channel used for detection of fluorescence with the wavelength 450 ± 50 nm. Place the cell population of unstained live cells (negative control) in the middle of the first "decade" of the logarithmic scale and check the settings obtained previously with the positive control. Using SYTO 16 green-stained sample of cells, adjust the voltage for the channel used for detection of fluorescence with the wavelength 525 ± 40 nm. Place the population of live cells with SYTO 16 green bright fluorescence at the border of the last and the penultimate "decades" of the logarithmic scale. Using DRAQ 7-stained cells, adjust the voltage for the channel used for detection of fluorescence with the wavelength 725 ± 20 nm. Place the population of unstained cells in the middle of the first "decade" of the logarithmic scale.
    (c) Check the voltages for fluorescent channels using samples of FMO-controls. FMO controls help to determine gating boundaries (*see* **Note 8**).
    (d) Bivariate dot plot "PO-PRO-1 versus DRAQ7" reveals three distinct subsets: viable cells remain impermeable to both dyes (gate "Viable" on dot plot A and B, Fig. 1); early apoptotic cells are permeable for PO-PRO-1, but they are still impermeable for DRAQ7 (gate "Po-Pro-1 + DRAQ7–" on dot plot A and B, Fig. 1); and late apoptotic/necrotic cells can be subsequently labeled with both DNA binding dyes (gate "Po-Pro-1 + DRAQ7+" on dot plot A and B, Fig. 1).
    (e) Bivariate dot plot "Syto 16 green versus DRAQ7" reveals three distinct subsets: viable cells effectively include Syto 16 green and are impermeable for DRAQ7 (gate "Viable" on dot plot C and D, Fig. 1); early apoptotic cells are Syto 16 green "dim," but they are still impermeable for DRAQ7 (gate "S16dimDRAQ7—"on dot plot C and D, Fig. 1); and late apoptotic/necrotic cells progressively loose SYTO 16 green fluorescence and gain bright DRAQ7 staining (gate "S16dim-to-low DRAQ7+" on dot plot C and D, Fig. 1) (*see* **Note 9**).

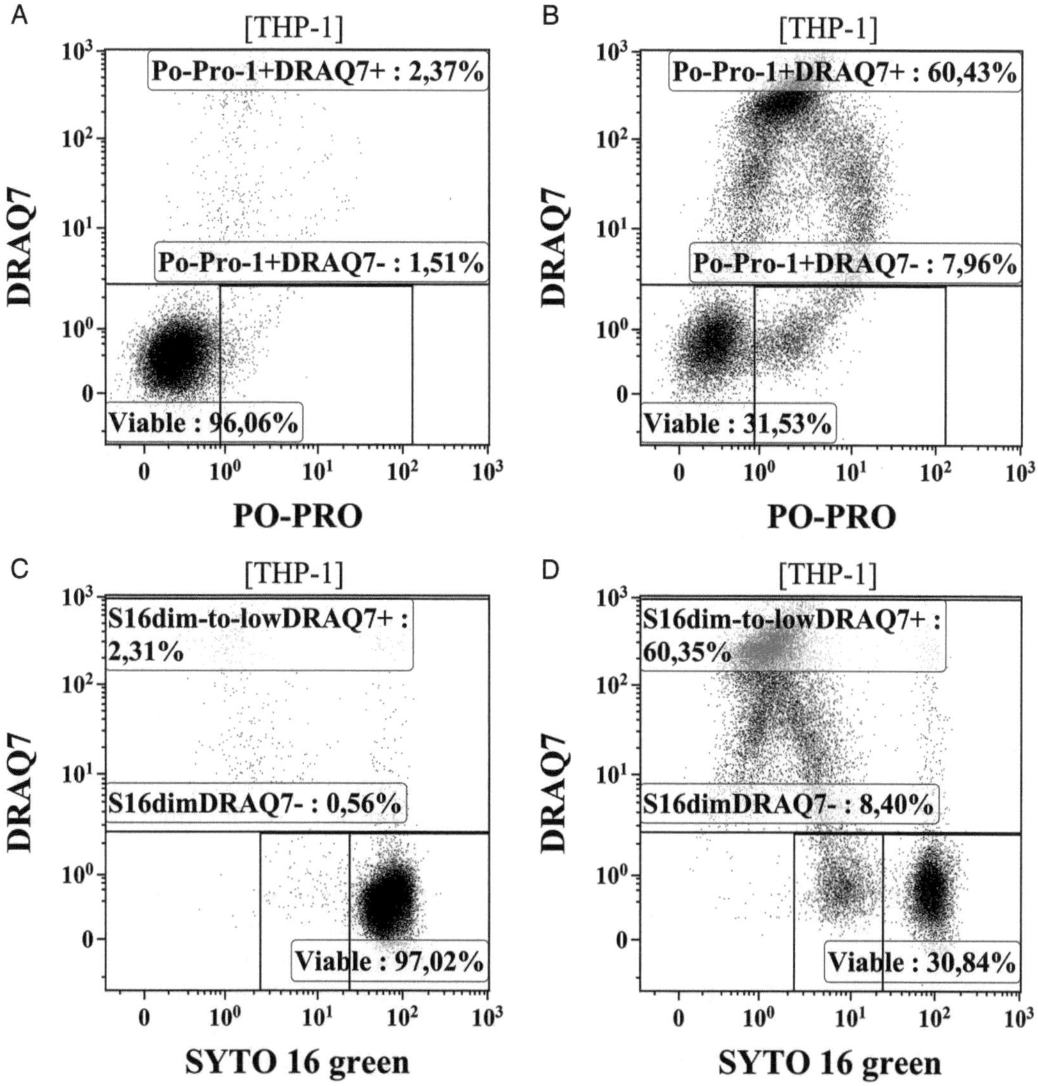

Fig. 1 Flow cytometry-based detection of early and late stages of apoptosis using three DNA-binding dyes. (**a**, **c**)—unexposed THP-1 cells. (**b**, **d**)—THP-1 cells exposed to 5 μM camptothecin. Cells were stained with PO-PRO-1 (**a** and **b**), or SYTO 16 green (**c** and **d**). All cells were additionally stained with DRAQ7. Viable, non-apoptotic/non-necrotic cells: gate "Viable" (**a**–**d**). Early stage of apoptosis cells: gate "Po-Pro-1 + DRAQ7–" (**a** and **b**), and gate "S16dimDRAQ7–" (**c** and **d**). These cells are DRAQ7-negative, but are permeable to PO-PRO-1 or have "dim" fluorescence of SYTO 16 green. Late stage of apoptosis and necrosis: gate "Po-Pro-1 + DRAQ7+" (**a** and **b**), and gate "S16dim-to-low DRAQ7+" (**c** and **d**). At the late stage of apoptosis and in necrosis the cells became DRAQ7 positive and lost the ability to bind SYTO 16

4 Notes

1. THP-1cells were cultured in RPMI 1640 medium supplemented with 10% fetal bovine serum (FBS), 100 units/mL penicillin, and 100 μg/mL streptomycin at 37 °C in a humidified

atmosphere with 5% CO_2. For experiments, cells were seeded in 96-well plates. To induce apoptosis, cells were treated for 24 h with 5 μM camptothecin.

2. Always use a sample of apoptotic cells as a positive control and a sample of untreated cells as a negative control. For the preparation of a positive control, use standard and well-known inducers of apoptosis—camptothecin, staurosporine, etc. Control cells should be processed in parallel with the investigated sample.
3. All stock solutions should be stored according to the manufacturer's instructions: working solutions of PO-PRO-1, SYTO 16 green and DRAQ7 are stable for at least 1 week if they are sterile, stored at 4 °C, and protected from direct light. It is recommended to use freshly prepared working solutions and protect them from direct light at all times.
4. Samples stained with fluorescent dyes are not photostable; therefore, store them in the dark before the analysis.
5. SYTO 16 green concentration should be optimized for each cell type. Although the final concentration of SYTO 16 green generally recommended for eukaryotic cells is 250 nM [5], in some cases SYTO 16 concentration as low as 4 nM was shown to be sufficient [6].
6. Wash step after staining with SYTO 16 green is important to obtain reproducible data and helps to discriminate between distinct cell subpopulations.
7. When only PO-PRO-1 and DRAQ7 are used (without SYTO 16 green), there is no need to wash samples prior to analysis. It saves time and is advantageous for analysis of multiple samples. Besides, it leaves the option to use 488 nm laser for excitation of fluorophores conjugated to antibodies for analysis of additional parameters of viable cell population.
8. Combining PO-PRO-1, SYTO 16 green and DRAQ7 does not require compensation, but, for each type of cells, FMO (fluorescence-minus-one) controls should be used.
9. Simultaneous application of PO-PRO-1 and SYTO 16 green with DRAQ7 provides an additional internal control for viable cells in each sample. In the case of PO-PRO-1 and DRAQ7 staining, live cells are double negative. This population of cells may also contain different DNA-free fragments including large protein aggregates, small apoptotic bodies, and others. In the case of SYTO 16 green and DRAQ7 staining, all live cells are SYTO 16 green-"bright." Therefore, the purity of different cell samples can be assessed by comparing their PO-PRO-1 and SYTO 16 dot plots. This is especially important when you are working with primary cell cultures or cells from different homogenized tissues.

References

1. Stokes L, Jiang LH, Alcaraz L et al (2006) Characterization of a selective and potent antagonist of human P2X(7) receptors, AZ11645373. Br J Pharmacol 149:880–887
2. Wlodkowic D, Skommer J (2007) SYTO probes: markers of apoptotic cell demise. Curr Protoc Cytom 7:7.33
3. Wlodkowic D, Skommer J, Darzynkiewicz Z (2012) Cytometry of apoptosis. Historical perspective and new advances. Exp Oncol 34:255–262
4. Akagi J, Kordon M, Zhao H et al (2013) Real-time cell viability assays using a new anthracycline derivative DRAQ7®. Cytometry A 83:227–234
5. Wlodkowic D, Skommer J, Pelkonen J (2007) Towards an understanding of apoptosis detection by SYTO dyes. Cytometry A 71:61–72
6. Sparrow RL, Tippett E (2005) Discrimination of live and early apoptotic mononuclear cells by the fluorescent SYTO 16 vital dye. J Immunol Methods 305:173–187

Chapter 9

Quick Detection of DNase II-Type Breaks in Formalin-Fixed Tissue Sections

Candace L. Minchew and Vladimir V. Didenko

Abstract

Blunt-ended DNase II-type breaks with 5′ hydroxyls are generated in phagocytic cells of any lineage during digestion of the engulfed DNA. These breaks indicate the ongoing active phagocytic reaction. They are produced by the acid deoxyribonuclease–DNase II which is the primary endonuclease responsible for DNA degradation after its engulfment.

Here, we present an express approach that detects blunt-ended 5′ OH DNA breaks in fixed tissue sections. The technique is simple to perform and takes only 60 min to complete. It can be useful in studies of the clearance of dying cells in oncological, inflammatory, and autoimmune disorders.

Key words Phagocytic digestion of DNA, Labeling of phagocytosis, Phagolysosomes, Clearance of cell death, Express detection of DNase II cleavage, 5′ OH DNA probes

1 Introduction

DNA breaks bearing 3′ phosphates and 5′ hydroxyls occur in phagolysosomes of those phagocytic cells that engulfed nuclear material from dying cells. They are produced by DNase II—a key endonuclease in the phagocytic degradation of DNA from apoptotic nuclei [1]. The amino acid sequence of this enzyme is highly conserved and close homologs of mammalian DNase II are present in invertebrates, such as worms and flies, which indicates its importance. The enzyme is located in lysosomes and is active in acidic conditions when it hydrolyzes the phosphodiester bonds in DNA [1].

Two different isoforms of lysosomal nucleases have been identified so far—DNase IIα and DNase IIβ [2–4]. The primary lysosomal enzyme is DNase IIα, often referred to as DNase II. This enzyme is expressed in all animal tissues. In contrast, DNase IIβ has a limited tissue distribution that varies between species. In mice it degrades nuclear DNA during lens cell differentiation [5] and is specifically expressed in liver [5], yet in human tissues it is absent from the liver but is highly expressed in the salivary gland [4].

Vladimir V. Didenko (ed.), *Fast Detection of DNA Damage: Methods and Protocols*, Methods in Molecular Biology, vol. 1644, DOI 10.1007/978-1-4939-7187-9_9, © Springer Science+Business Media LLC 2017

Overall, the ubiquitous expression pattern of DNase II (DNase IIα) and its presence in all tissues confirms that it is the principal lysosomal DNA digestion enzyme, whereas DNase IIβ performs specialized functions in selected tissues in different mammalian species and may also function as a secreted enzyme [1, 4].

Both isoforms of lysosomal DNase II produce DNA breaks that contain 3′ PO_4/5′ OH at the ends. DNA breaks having this configuration are often referred to as DNase II-type breaks [6–8]. This sort of DNA cleavage differs from DNA cuts produced by DNase I, DNase I-like nucleases, and caspase-activated deoxyribonuclease (CAD), which all create the reversed 3′ OH/5′ PO_4 end-group pattern. Such DNase I-type breaks are produced in apoptotic execution and are used as specific markers of apoptotic cells [9–11]. In the same way, the DNase II-type breaks serve as characteristic markers of DNase II activity and indicate the digestion of engulfed DNA [7]. Specific detection of these breaks in cells in fixed tissue sections indicates the ongoing active phagocytic reaction and labels phagocytes of any lineage participating in active clearance of dead cells [12].

Here, we present an express version of the technique which labels DNase II-type DNA breaks. The protocol takes only 60 min to complete. The increased speed of the assay is enabled by two factors: the quicker processing of sections, and the use of the ultrafast labeling enzyme—vaccinia topoisomerase (*see* **Note 1**).

The described approach selectively labels blunt-ended 5′ OH DNA breaks in formaldehyde-fixed, paraffin-embedded tissues. The detected type of DNA breaks localize in phagolysosomes of phagocytizing cells and are produced by DNase II. The brightly labeled phagolysosomes filled with semi-digested DNA can be easily identified in the cytoplasm of phagocytic cells under fluorescence microscope observation (Fig. 1).

The technique does not label single-stranded DNA breaks or DNase I-type cleavage, such as the caspase-initiated apoptotic DNA fragmentation producing DNA breaks with 5′ PO_4 instead of 5′ OH. Instead, the labeling indicates the active phagocytic clearance of dying (either apoptotic or necrotic) cells. The assay is simple, economic, and fast. It can be easily mastered by a researcher new to the field of in situ labeling (*see* **Note 2**).

The assay utilizes the unique enzymatic properties of vaccinia DNA topoisomerase I (TOPO), a virus-encoded eukaryotic type IB topoisomerase. When applied to tissue sections this enzyme specifically attaches the preactivated blunt-ended hairpin-shaped oligoprobes to the 5′ OH ends of blunt-ended DNA breaks. This type of ligation is not possible for DNA ligases which all require 5′ PO_4 at DNA ends. However, vaccinia topoisomerase I used in this assay joins DNA molecules employing a different mechanism.

In nature this topoisomerase untwists cellular DNA to release its torsional stress. At first the enzyme binds to the specific

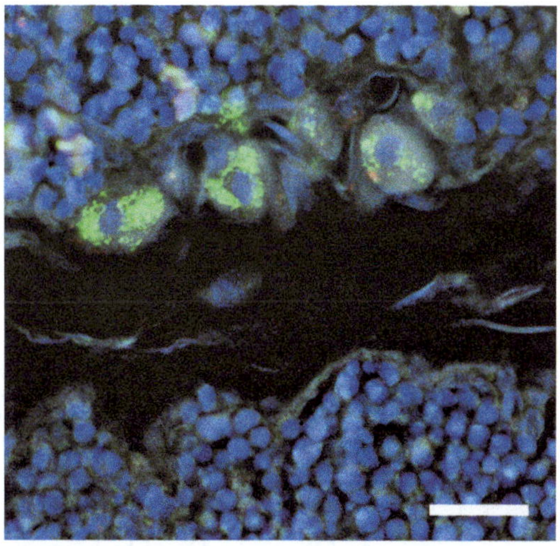

Fig. 1 Phagolysosomes in macrophages clearing apoptotic cells. Phagolysosomes in the cytosol of macrophages, which engulfed apoptotic cells, are brightly labeled by the TOPO assay (FITC—green fluorescence). Nuclei of macrophages are visualized by DAPI (blue fluorescence). Rat thymus 24 h after injection of dexamethasone [7, 12]. Bar, 50 μm

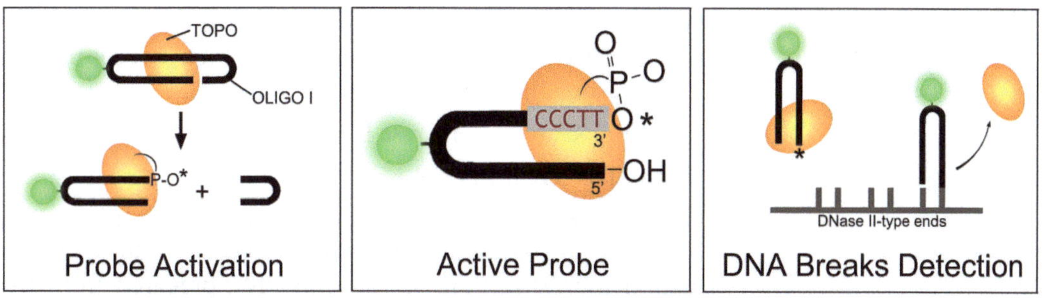

Fig. 2 Assay for detection of DNase II-type breaks in tissue sections. *Left:* TOPO binds to the recognition sequence in the double-hairpin oligo and cleaves it activating left hairpin for ligation to 5′ OH DNA termini. *Center:* Active Probe—fluorescent complex of oligo and TOPO covalently linked at the ligatable 3′ end of the hairpin (*asterisk*). *Right:* Active probe specifically ligates to the 5′ OH double stand DNase II-type break generated by DNase II in phagocytic cells digesting engulfed chromatin

recognition sequence and makes a single-strand cut at its end thus freeing an opposite 5′ OH DNA terminus. The DNA molecule then rotates around the remaining strand and releases the stress. Next, TOPO re-ligates the DNA strand back to the momentarily released 5′ OH DNA terminus. The assay uses this re-ligation activity of the topoisomerase to detect DNase II breaks with 5′ OH ends. Its labeling principle is presented in Fig. 2 (*see* **Note 1**).

This chapter presents a step-by-step express labeling protocol which requires 1 h to complete. The protocol deals with the rapid preparation and usage of fluorescent TOPO probes in fixed tissue sections to label the phagocytic clearance of dying cells.

2 Materials

1. 5–6 μm-thick sections cut from paraformaldehyde-fixed, paraffin-embedded tissue blocks. Sections of different thicknesses (3–50 μm-thick) can potentially be used but might require different time of dewaxing and/or Proteinase K treatment (**step #3**). Use charged and precleared slides that retain sections well.

   For the positive control experiments, as the source of DNase II type DNA breaks, we recommend using sections of dexamethasone-treated rat thymus (*see* **Note 3**).
2. Xylene.
3. 70, 80, and 96% Ethanol.
4. Oligo 1. Double-hairpin vaccinia topoisomerase I cleavable oligonucleotide. The oligonucleotide is labeled with a single fluorescein.

   5′-AAGGGACCTGC**F**GCAGGTCCCTTAACGCAT ATGCGTT-3′;

   F – FITC-dT

   PAGE or HPLC purification is recommended. Dilute with bidistilled water to 100 pmol/μL stock concentration. Store at −20 °C protected from light.
5. Vaccinia DNA topoisomerase l–6 pmol/μL stock (*see* **Note 4**).
6. 50 mM Tris–HCl, PH 7.4.
7. Proteinase K (Roche Diagnostics Corporation, Inianapolis, IN) 20 mg/mL stock in distilled water. Store at −20 °C. In the reaction use 50 μg/mL solution in PBS, prepared from the stock. Do not reuse (*see* **Note 5**).
8. Vectashield with DAPI (Vector Laboratories, Burlingame, CA).
9. Phosphate-buffered saline (1× PBS): dissolve 9 g NaCl, 2.76 g $NaH_2PO_4{\cdot}H_2O$, 5.56 g $Na_2HPO_4{\cdot}7H_2O$ in 800 mL of distilled water. Adjust to pH 7.4 with NaOH, and fill to 1 L with distilled water.
10. Fluorescent microscope with appropriate filters and objectives.

3 Method

3.1 Labeling 5′ OH Blunt-Ended DNA Breaks in Tissue Sections

1. Place the sections in a slide rack and dewax in xylene for 5 min, ransfer to a fresh xylene bath two more times for an additional 2 min each. For each transfer, dip cassette up and down three times (*see* **Note 6**).

2. Rehydrate by passing through graded ethanol concentrations: 96% Ethanol—2 × 2 min; 80% Ethanol—2 × 2 min; 70% Ethanol—2 min; water—2 × 2 min. For each transfer, dip cassette up and down three times (*see* **Note 6**).
3. Digest section with Proteinase K. Use 100 μL of a 50 μg/mL solution per section. Incubate for 10 min at room temperature (23 ° C) in a humidified chamber (*see* **Note 7**).
4. Rinse in distilled water for 3 × 2 min. For each transfer, dip cassette up and down three times (*see* **Note 6**).
5. While sections are rinsing, combine 100 pmoles of Probe 1 and 100 pmoles (3.3 μg) of TOPO in a solution of 50 mM Tris–HCL, pH 7.4 (*see* **Note 8**). Use 100 μL of this reaction solution per section.
6. Aspirate water from sections and apply the labeling mix containing the oligoprobes and TOPO enzyme.
7. Incubate for 15 min at room temperature (23 °C) in a humidified chamber, protected from light.
8. Wash sections 3 × 2 min in distilled water. For each transfer, dip cassette up and down three times (*see* **Note 6**).
9. Cover sections with an antifading solution (Vectashield with DAPI), coverslip and analyze the signal using a fluorescent microscope. Double-strand DNA breaks with 5′ OH will fluoresce green.

4 Notes

1. TOPO can perform the specific ligation to 5′ OH ends in tissue sections in 15 min [13]. This high speed of attachment is explained by the intrinsically fast ligation activity of the vaccinia topoisomerase enzyme. In the kinetic analysis of DNA strand cleavage and ligation reactions, this topoisomerase demonstrated ligation of 85% of oligoprobes within a 15 s interval [14].

   The detailed analysis of various probe designs and this labeling approach that uses vaccinia topoisomerase are discussed elsewhere [7, 13].
2. The other assays commonly used for detection of fragmented DNA in fixed cells rely on labeling of 3′ OH groups (TUNNEL assay) or 5′ PO_4 groups (in situ ligation) and cannot label the same marker [7–9].
3. Apoptotic thymus contains both DNase I type and DNase II type DNA breaks [15] and is useful in control experiments. To make apoptotic thymus, subcutaneously inject Sprague-Dawley rats (150 g) with 6 mg/kg dexamethasone (Sigma) dissolved

in 30% dimethyl sulfoxide in water. Animals should be sacrificed 24 h post injection. To fix the thymus, incubate it for 18 h in 4% paraformaldehyde. For paraffin embedding, pass the tissue through graded alcohols finishing with 100% ethanol, then place it overnight in chloroform and embed in paraffin.

4. Highly concentrated TOPO, which works well with the described assay, can be purchased from Millipore, sold as a part of the ApopTag® ISOL Dual Fluorescence Apoptosis Detection Kit. We have also used the highly concentrated preparation of this enzyme obtained from Vivid Technologies (Houston, TX).
5. At concentrations higher than 1 mg/mL proteinase K is very stable and can be stored for years at −20 °C. At low concentrations (~10 μg/mL) it is less stable and its activity gradually decreases due to autolysis [16].
6. We consider the cassette dipping important for speeding up the processing of sections. The gentle and consistent dipping enhances the interaction between the section and the dewaxing (or washing) solution due to the increased convection.
7. The duration of Proteinase K digestion may need adjustment depending on the tissue type. Harder tissues might require longer digestion. Times of 10 min are usually used. The complete omission of the digestion step results in a weaker signal. On the other hand, overdigestion can result in signal disappearance and disruption of entire section.
8. In the initial experiments we used 215 pmoles (7.1 μg) of the enzyme per section in 25 μL of the reaction mix. However, the topoisomerase concentration can be significantly reduced without any loss of sensitivity. We later used four times less of the enzyme per section 53 pmol (1.76 μg per section) with similar results. Reducing the amount of enzyme to 26 pmol (880 ng per section) resulted in a weaker signal and 266 fmol (8.8 ng per section) of enzyme produced no signal.

Acknowledgment

This research was supported by grant R01 NS082553 from the National Institute of Neurological Disorders and Stroke, National Institutes of Health and by grants R21 CA178965 from the National Cancer Institute, National Institutes of Health and R21 AR066931 National Institute of Arthritis and Musculoskeletal and Skin Diseases, National Institutes of Health (all to V.V.D.).

References

1. Evans CJ, Aguilera RJ (2003) DNase II: genes, enzymes and function. Gene 322:1–15
2. Shiokawa D, Tanuma SI (2001) Isolation and characterization of the DLAD/Dlad genes, which lie head-to-head with the genes for urate oxidase. Biochem Biophys Res Commun 288(5):1119–1128
3. Shiokawa D, Tanuma SI (1999) DLAD, a novel mammalian divalent cation-independent endonuclease with homology to DNase II. Nucleic Acids Res 27(20):4083–4089
4. Krieser RJ, MacLea KS, Park JP, Eastman A (2001) The cloning, genomic structure, localization, and expression of human deoxyribonuclease IIbeta. Gene 269:205–216
5. Nishimoto S, Kawane K, Watanabe-Fukunaga R, Fukuyama H, Ohsawa Y et al (2003) Nuclear cataract caused by a lack of DNA degradation in the mouse eye lens. Nature 424:1071–1074
6. Didenko VV (2011) 5'OH DNA breaks in apoptosis and their labeling by topoisomerase-based approach. Methods Mol Biol 682:77–87
7. Minchew CL, Didenko VV (2011) Fluorescent probes detecting the phagocytic phase of apoptosis: enzyme-substrate complexes of topoisomerase and DNA. Molecules 16:4599–4614
8. Hsiang Y, Huey-Jen L, Tai-Wei L, Szecheng JL (2015) Autonomous and non-autonomous roles of DNase II during cell death in *C. elegans* embryos. Biosci Rep 35:e00203. doi:10.1042/BSR20150055
9. Hornsby PJ, Didenko VV (2011) In situ ligation: a decade and a half of experience. Methods Mol Biol 682:49–63. doi:10.1007/978-1-60327-409-8_5
10. Didenko VV (2011) In situ ligation simplified: using PCR fragments for detection of double-strand DNA breaks in tissue sections. Methods Mol Biol 682:65–75. doi:10.1007/978-1-60327-409-8_6
11. Minchew CL, Didenko VV (2012) In vitro assembly of semi-artificial molecular machine and its use for detection of DNA damage. J Vis Exp 59:e3628
12. Minchew CL, Didenko VV (2014) Assessing phagocytic clearance of cell death in experimental stroke by ligatable fluorescent probes. J Vis Exp 87:e51261. doi:10.3791/51261
13. Minchew CL, Didenko VV (2014) Nanoblinker: Brownian motion powered bio-nanomachine for FRET detection of phagocytic phase of apoptosis. PLoS One 9:e108734. doi:10.1371/journal.pone.0108734
14. Stivers JT, Shuman S, Mildvan AS (1994) Vaccinia DNA topoisomerase I: single-turnover and steady-state kinetic analysis of the DNA strand cleavage and ligation reactions. Biochemistry 33:327–339
15. Didenko VV, Minchew CL, Shuman S, Baskin DS (2004) Semi-artificial fluorescent molecular machine for DNA damage detection. Nano Lett 12:2461–2466
16. Sweeney PJ, Walker JM (1993) Proteinase K (EC 3.4.21.14). Methods Mol Biol 16:305–311

Part II

Rapid Detection (2–3 h)

Chapter 10

Express γ-H2AX Immunocytochemical Detection of DNA Damage

Nate Hopp, Jodi Hagen, Birte Aggeler, and Alexander E. Kalyuzhny

Abstract

DNA can be damaged by many environmental factors including chemical agents and ionizing radiation which induce the formation of DNA double-stranded breaks (DSBs). If DSBs are not repaired in a timely fashion this may cause the disruption of genome integrity, which can result in cancer development. Typically, DSBs are followed by phosphorylation of histone protein H2AX, a member of the H2A family. Immunocytochemical detection of phosphorylated H2AX (e.g., γ-H2AX) appears to be a useful technique for assessing DNA damage. Such an assessment is easy to do by analyzing labeling for γ-H2AX under the microscope and does not require an expensive laboratory setup. Using HeLa cells treated with camptothecin as a model, we developed an easy-to-run protocol to analyze DSBs. Our protocol can be applied to testing the potency of different chemicals to induce DSBs in different types of cells and requires around 2 h to complete.

Key words HeLa, γ-H2AX, Phosphorylated H2AX, Camptothecin, Double-strand DNA breaks

1 Introduction

Histones are highly conserved alkaline proteins that make up the core structure of the nucleosome. There are five families of histones: H1/H5, H2A, H2B, H3, and H4. The core histones form an octamer comprised of a pair of H2A-H2B dimers and a H3-H4 tetramer [1]. 147 DNA base pairs wrap around the octamer 1.65 times in a left-handed super helical turn to form the nucleosome [2]. In eukaryotes the nucleosome is the fundamental subunit in the chromatin, in which DNA is wound up in tight spools. This organizational structure allows for the 1.8 m of human DNA to condense to about 90 μm.

Histones were first isolated in the 1884 by Albrecht Kossel, who characterized nucleic acids. For many years, histones were thought to just be a storage platform to condense the chromatin. That theory began to change in 1964 when Murray reported lysine methylation modifications on isolated histones from calf and rabbit

Vladimir V. Didenko (ed.), *Fast Detection of DNA Damage: Methods and Protocols*, Methods in Molecular Biology, vol. 1644, DOI 10.1007/978-1-4939-7187-9_10, © Springer Science+Business Media LLC 2017

thymus [3]. This discovery sparked the need to better understand the structure of histones. Later in the 1980s, it became clear that deletions and modifications to the core histones can also suppress gene activation [4]. Today, we understand that histones undergo numerous modifications that can affect many cellular activities, such as gene regulation, DNA repair, mitosis, and meiosis [5, 6]. Each type of histone has its own small subset of variants that differ slightly in their amino acid sequence, mostly in the N-terminus. There are currently over 50 known histone variants [7].

These variants and subsequent modifications are beginning to be well characterized, but understanding the downstream activity and effects of these unique histones are mostly still a mystery [8]. Modifications include lysine methylation, arginine methylation, arginine citrullination, lysine acetylation, and the phosphorylation of serine, threonine and tyrosine residues. One such modification is the phosphorylation of a serine at position 139 (human) of the variant H2AX in response to double-stranded DNA breaks (DSBs) [9]. Phosphorylated H2AX (referred to as γ-H2AX) is created very close to the site of a DSB and is thought to play a pivotal role as a recruiter for the DSB repair machinery [10]. The repair sites of DSBs, known as "foci," can be visualized as nuclear aggregates under fluorescent microscopy. This led to phosphorylated γ-H2AX becoming a valuable and widely used biomarker in research.

With the vast set of conditions and compounds that can induce or repair DBS breaks there is a strong need for a rapid test. Thanks to the development of highly specific antibodies for immunocytochemistry, it is now possible to complete tests in a short period of time and analyze the staining results right away under the microscope.

In our study, we analyzed response of HeLa cells to camptothecin treatment followed by immunofluorescence labeling for the detection of γ-H2AX.

2 Materials

2.1 Cell Culture

1. HeLa Cells: Human Epithelial Cervix Adenocarcinoma, tenth passage (American Type Culture Collection (ATCC), Manassas, VA).
2. T75 flasks.
3. 1× Phosphate-buffered saline (PBS), pH 7.4.
4. Eagle's Minimum Essential Medium (EMEM) supplemented with 10% Fetal Bovine Serum and 1% Penicillin/Streptomycin.
5. 1× Trypsin EDTA reagent: Prepare a working solution of 1× Trypsin EDTA by diluting 100 mL of 10× Trypsin EDTA into 900 mL of sterile deionized H_2O.
6. Centrifuge allowing spinning 50 mL culture tubes at 500 × *g*.

7. Hemacytometer to count lymphocytes under the microscope.
8. Trypan Blue Dye to evaluate cell viability.
9. Upright microscope equipped with bright-field illumination and phase contrast condenser.

2.2 Analysis of DSBs

1. Reagent to induce DNA breaks: Prepare a 2 mg/ mL stock solution of camptothecin in tissue culture grade dimethyl sulfoxide by adding 2 mg of camptothecin in 1 mL of DMSO and store at −20 °C.
2. Cell fixative: 4% Paraformaldehyde in 0.2 M Sorenson's Phosphate buffer made by mixing 0.2 M NaH_2PO_4 with 0.2 M Na_2HPO_4 (4% PFA).
3. Antibody Diluent reagent: 1× PBS, 0.3% Triton, 1% Normal Donkey serum, 1% Bovine Serum Albumin, and 0.01% Sodium Azide.
4. Human Phospho-Histone H2AX (S139) Antibody (R&D Systems, Minneapolis, MN).
5. Donkey Anti-Rabbit IgG Northern Lights™ NL557-conjugated Antibody (R&D Systems, Minneapolis, MN).
6. Northern Lights Guard Mounting Media (R&D Systems, Minneapolis, MN).
7. DAPI to counterstain nuclei.
8. Conventional fluorescence or laser scanning confocal microscope.

3 Methods

3.1 Cell Culture

1. Culture HeLa cells in EMEM culture media to a population around 1--2 × 10^6 cells in a 37 °C/CO_2 humidified incubator (*see* **Note 1**).
2. In a sterile hood pour out old culture media and gently rinse the cell layer with sterile PBS three times.
3. Add 3 mL of 1× Trypsin–EDTA over the cell layer and place the flask in a 37 °C/CO_2 humidified incubator for 5 min (*see* **Note 2**).
4. Once the cells become suspended, take the cells to a sterile hood and gently aspirate the cells into a 50 mL culture tube. Add 7 mL of EMEM to the tube and place in centrifuge.
5. Centrifuge the 50 mL culture tube containing the cells at 500 × *g* for 5 min.
6. Take the cells to a sterile hood and gently pour out the supernatant. Resuspend the pellet by adding 2 mL of EMEM to the tube and gently mix.

7. Mix 30 μL of cells with 30 μL of Trypan Blue in a small vial and pipette 10 μL of mixture onto a hemacytometer with a coverslip. Don't add more than 10 μL.
8. Count the cells on the hemacytometer under a bright field microscope (**Note 3**). Cells that were stained with Tripan Blue are considered dead cells. By counting live and dead cells it is posible to determine a percentage of live cells using this formula: % live cells = (Tripan Blue stained cells/total number of cells) × 100. For this experiment it is recommended to maintain cell viability around 90–95%.
9. Dilute your cells to a density of 5×10^4 cells per 1 mL of EMEM.
10. Pipette 125 μL of HeLa cells per well onto a 96-well CellCarrier glass-bottom plate. Cover and incubate overnight in a 37 °C/CO_2 humidified incubator.

3.2 Induction of DNA Breaks by Iinhibiting DNA Religation

1. Calculate the working volume of reagent needed to induce DNA breaks. Prepare 10 μM working solution of camptothecin in EMEM by adding 1.74 μL of stock camptothecin per 1 mL of EMEM. Prepare serial dilutions of camptothecin at different concentrations into EMEM to achieve working solutions.
2. Transfer the plate with HeLa cells from incubator into a sterile hood. Remove the cover and aspirate the culture media from well(s) that are to be treated and discard. Replace with 125 μL of working camptothecin solution to well(s) and incubate in a 37 °C/CO_2 humidified incubator (*see* **Note 4**).
3. Once the final incubation time(s) has been reached, the cells will need to be fixed. Add 125 μL of the 4% PFA solution to each well with culture media to achieve a 1:1 ratio of culture media to 4% PFA. Cover and incubate the plate for 20 min at room temperature.

3.3 Immunocytochemical Detection of DNA.

1. Wash the plate three times with PBS.
2. Make a working solution of the primary antibody anti- human S139 phosphorylated H2AX protein (aka γ-H2AX) in the antibody diluent buffer at a concentration of 1 μg/mL.
3. Add 100 μL of the diluted anti-γ-H2AX antibody solution to each well. Cover and incubate the plate for 1 h at room temperature.
4. Wash plate three times with PBS.
5. Make a working solution of the fluorescent secondary Donkey Anti-Rabbit IgG NorthernLights™ NL557-conjugated antibody by diluting 1:200 in the antibody diluent buffer.

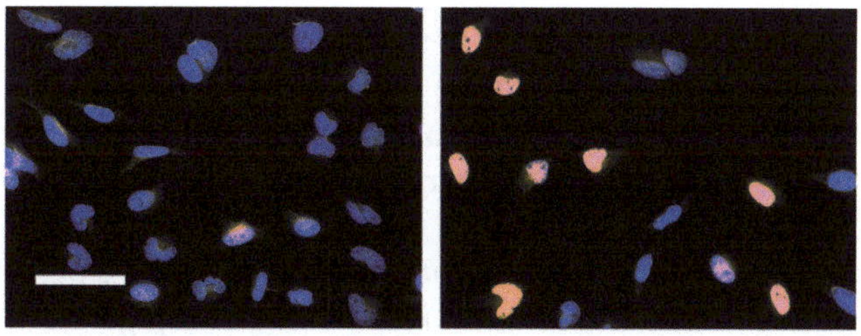

Fig. 1 Confocal microscopy images of γ-H2AX labeling in the nuclei of HeLa cells untreated (control) and treated for 60 min with 1 μM camptothecin (red fluorescence). Cell nuclei were counterstained with DAPI (blue fluorescence). Note a robust effect of camptothecin on inducing DSBs. Scale bar = 20 μm

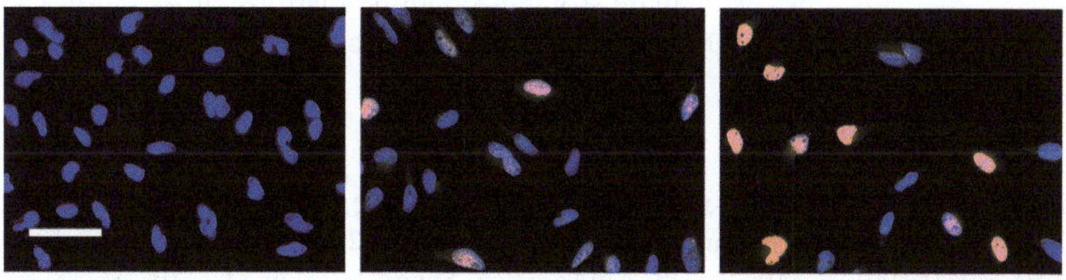

Fig. 2 Confocal microscopy images of γ-H2AX labeling in the nuclei of HeLa cells treated for different periods of time with 1 μM camptothecin (red fluorescence). Cell nuclei were counterstained with DAPI (blue fluorescence). Note an increase in γ-H2AX labeling intensity with increasing duration of camptothecin treatment. Scale bar = 20 μm

6. Add 100 μL of the diluted fluorescent secondary antibody to each well. Cover and incubate the plate at room temperature for 30 min.
7. Wash plate three times with PBS.
8. To each well add 50 μL of DAPI mixed 1:10,000 in Northern Lights Guard Mounting media and cover with the plate sealer (*see* **Note 5**) (Figs. 1 and 2).

4 Notes

1. Use T75 flasks and follow ATCC recommendations for culturing and subdividing HeLa cells.
2. This treatment is designed to gently release cells from the surface and disrupt cell-cell bonds to create a cell suspension by using the proteolytic enzyme Trypsin. The 1× Trypsin–EDTA can take up to 15 min to suspend an adherent cell line, such as

HeLa. Inspect the flask to make sure cells are starting to detach after 5 min of incubation.

3. Locate the center of the hemacytometer under the microscope. There is a grid of 25 even squares. Count the total number of cells in 10 of the squares and calculate the average cells per square and multiply by 25 (total number of squares in that area) and then multiply by 2 (cell dilution factor), and then multiply by 10,000 to determine the number of cells in 1 mL of original cell suspension.
4. When performing cell stimulations with multiple time intervals, it is recommended to start with the longest time (180 min) and end with the shortest time (15 min). This allows for all stimulations to end at the same time and immediately proceed to the cell fixation step.
5. Assay plate storage recommendations: at 4 °C for a week and −20 °C for long-term storage.

References

1. Eickbush TH, Moudrianakis EN (1978) The histone core complex: an octamer assembled by two sets of protein-protein interactions. Biochemistry 17:4955–4964
2. Luger K, Mäder AW, Richmond RK et al (1997) Crystal structure of the nucleosome core particle at 2.8 aA resolution. Nature 389:251–260
3. Murray K (1964) The occurrence of epsilon-N-methyl lysine in histones. Biochemistry 3:10–15
4. Kayne PS, Kim UJ, Han M et al (1988) Extremely conserved histone H4 N terminus is dispensable for growth but essential for repressing the silent mating loci in yeast. Cell 55:27–39
5. Kaufman-Szymczyk A, Majewski G, Lubecka-Pietruszewska K et al (2015) The role of sulforaphane in epigenetic mechanisms, including interdependence between histone modification and DNA methylation. Int J Mol Sci 2015: 29732–29743
6. Song N, Liu J, An S et al (2011) Immunohistochemical analysis of histone H3 modifications in germ cells during mouse spermatogenesis. Acta Histochem CytochemActa Histochemica et Cytochemica 44:183–190
7. Cheema M, Ausió J (2015) The structural determinants behind the epigenetic role of histone variants. Genes 6:685–713
8. Huang J, Marco E, Pinello L et al (2015) Predicting chromatin organization using histone marks. Genome Biol 16:162
9. Turinetto V, Giachino C (2015) Multiple facets of histone variant H2AX: a DNA double-strand-break marker with several biological functions. Nucleic Acids Res 43:2489–2498
10. Chapman JR, Taylor MR, Boulton SJ (2012) Playing the end game: DNA double-strand break repair pathway choice. Mol Cell 47:497–510

Chapter 11

Rapid Detection of γ-H2AX by Flow Cytometry in Cultured Mammalian Cells

Denis Firsanov, Liudmila Solovjeva, Olga Lublinskaya, Valeriy Zenin, Igor Kudryavtsev, Maria Serebryakova, and Maria Svetlova

Abstract

Methods commonly used for detection of DNA double-strand breaks (DSBs) and analysis of cell death are generally time-consuming, and, therefore, any improvements in these techniques are important for researchers and clinicians. At present, flow cytometry is the most rapid method for detection of DSBs and cell viability. In this chapter, we provide our experience and methodological modification of flow cytometry protocol for the detection of γ-H2AX, a well-known marker of DSBs, in fixed mammalian fibroblasts. The modifications permit a reduction in the time required for DSB detection by flow cytometry.

Key words Flow cytometry, γ-H2AX, DNA double-strand breaks, Rapid detection, Fibroblasts

1 Introduction

Living organisms are exposed to different intrinsic and extrinsic stresses that damage DNA and can induce programmed cell death or apoptosis. DNA double-strand breaks (DSBs) represent the most harmful type of DNA damage which is accompanied by phosphorylation of histone H2AX (producing γ-H2AX) in the vicinity of broken DNA ends [1]. It has been shown that the number of DSBs corresponds to the number of γ-H2AX focal sites in cell nuclei [2]. At present, γ-H2AX is widely used as a marker of DSBs in different fields of experimental research and clinical practice for estimation of DNA damage and repair.

Flow cytometry assay is a fast and reliable method for the estimation of γ-H2AX level in damaged cells grown in culture or obtained from peripheral blood of patients. It gives possibility to estimate DNA damage in a large number of cells using an antibody to γ-H2AX, and monitor cell cycle distribution in populations of cells.

In this chapter, we present two accelerated procedures that use flow cytometry for the detection of γ-H2AX in fixed cells. The first

Vladimir V. Didenko (ed.), *Fast Detection of DNA Damage: Methods and Protocols*, Methods in Molecular Biology, vol. 1644, DOI 10.1007/978-1-4939-7187-9_11,

approach—Protocol 1 (*see* Subheading 3.1) is a speeded up modification of the BD Pharmingen™ flow cytometry protocol. The second technique—Protocol 2 (*see* Subheading 3.2) uses formaldehyde fixation and the alternative permeabilizing agent saponin.

1.1 Protocol 1: Accelerated Detection of γ-H2AX by Modified BD Pharmingen™ Procedure

At present, the most reliable reagents for flow cytometry are provided by BD Pharmingen company. Here, we describe a time-saving modification of flow cytometry assay for γ-H2AX detection in fixed mammalian fibroblasts based on the original intracellular staining protocol proposed by BD Pharmingen™ [3]. We use the fixation/permeabilization solution provided by this company and direct- or indirect immunochemical γ-H2AX detection using Alexa Fluor 488-conjugated anti-γ-H2AX antibody, or primary nonlabeled anti-γ-H2AX antibody and secondary fluorophore-conjugated antibodies, while DNA is counterstained with DAPI. Overall processing time is minimized. The complete procedure takes only 2 h 15 min (compared to 2 h 40 min in case of using standard BD Pharmingen protocol) due to incubation of cells with antibodies at 37 °C instead of room temperature.

1.2 Protocol 2: Use of Saponin-Based Permeabilization in Detection of γ-H2AX

An alternative protocol for γ-H2AX detection provided in this chapter is based on another kind of treatment: formaldehyde fixation and permeabilization with a mild detergent saponin. Saponins are the mixture of glycosides with soap-like foaming characteristics. Saponins permeabilize cells due to interaction with cholesterol in cell membrane and are often utilized for phospho-epitope staining protocols [4]. As the result of the breakage of association of cholesterol with phospholipids, the membrane becomes permeable [5]. Gentle permeabilization by saponin does not destroy fibroblasts, it is enough to detect γ-H2AX phospho-epitopes, and the percentage of γ-H2AX-positive cells after irradiation does not differ from that obtained by the application of the first protocol. The processing time is 2 h 15 min.

Protocol 1 is preferable because it gives well-repeatable results and minimal loss of cells during the procedure. The cell loss while using protocol 2 is more significant; however, this protocol is more cost-effective because it does not need BD Pharmingen reagents.

Flow cytometry protocol for staining of mononuclear blood cells, described by other researchers, does not include the stage of fixation, but, nevertheless, requires 4–6 h [6].

2 Materials

2.1 Materials for Cultivation of Primary Fibroblasts

1. Minimum essential medium (MEM).
2. Fetal calf serum (FCS).
3. 100× antibiotic stock solution, 10,000 U/mL penicillin, 10,000 μg/mL streptomycin (Invitrogen).

4. Plastic flasks with a surface area of 25 cm^2.
5. Calcium and magnesium-free phosphate-buffered saline (PBS), pH 7.4 (*see* **Note 1**).
6. Versene solution: 0.2 g EDTA(Na4) per liter of calcium and magnesium-free PBS. Versene is an EDTA solution used for the gentle non-enzymatic cell dissociation.
7. 0.25% Trypsin–EDTA solution for detachment of adherent cells in culture.
8. CO_2 incubator.

2.2 Materials for Cell Fixation/Permeabilization

1. Vortex mixer.
2. Refrigerated laboratory centrifuge.
3. 1.5 mL Eppendorf tubes.
4. 5 mL polystyrene tubes for flow cytometry.
5. Transcription Factor Buffer Set for flow cytometry, cat #562574 (BD Pharmingen) containing: TF Perm/Wash buffer, TF Diluent Buffer, TF Fix/Perm buffer (*see* **Note 2**).
6. Double-distilled water.
7. Antibodies: Mouse monoclonal anti-phospho-histone H2AX antibody, cat # 05-636 (Millipore), secondary goat anti-mouse IgG (H + L) Alexa Fluor 488-conjugated antibody (Invitrogen), or, alternatively, mouse monoclonal anti- phospho-histone H2AX (Ser139) Alexa Fluor488-conjugated antibody, cat # 05-636AF488 (Millipore).
8. 1% saponin stock solution in ddH_2O.
9. 10% Triton X-100 stock solution in ddH_2O.
10. 4′, 6-diamidino-2-phenylindole (DAPI) 100 μg/mL solution in PBS.
11. Flow Cytometer, such as Beckman Coulter CytoFlex Cytometer or similar.

3 Methods

Both protocols are based on our experience of γ-H2AX detection in irradiated primary human fibroblasts and Syrian hamster fibroblasts isolated from skin of newborn animals.

3.1 Protocol 1. Modified BD Pharmingen™ Procedure for γ-H2AX Detection

In this protocol, we use reagents from BD Pharmingen. An improved technique takes a little more than 2 h excluding the time needed for trypsinization and harvesting the cells (Fig. 1).

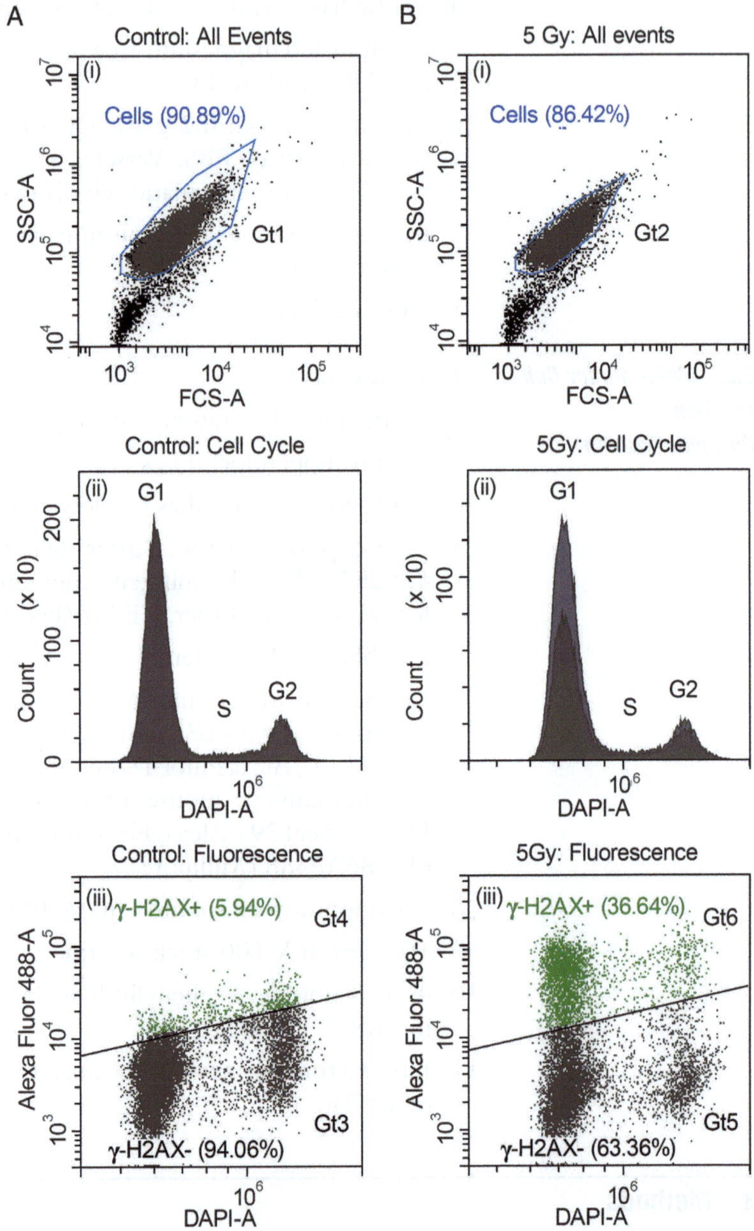

Fig. 1 γ-H2AX detection by flow cytometry in human embryonic fibroblasts. (**a**) Unirradiated cells, (**b**) cells irradiated at the dose of 5 Gy. 1 h after irradiation, cells were processed for flow cytometry analysis using Protocol 1 (Subheading 3.1). (*i*) Viable cells are gated (Gt1, Gt2) on the scatter plot FSC versus SSC in (**a**) and (**b**) for exclusion of cell debris. (*ii*) The distribution of cells in the phases (G1, S, G2) of the cell cycle is shown on the histogram plot 405 nm laser area (DAPI-A) versus cell count in (**a**) and (**b**). (*iii*) On the bivariate dot plot DAPI-A versus 488 nm laser area (Alexa Fluor 488-A), γ-H2AX-positive cells depicted in *green* (Gt4) and γ-H2AX-negative cells depicted in *black* (Gt3) are gated in control cell population (**a**). The same gating is applied for irradiated cells (**b**). Gt5 on the dot plot for irradiated cells (**b**) represents γ-H2AX-negative cell population with background staining with goat anti-mouse Alexa Fluor 488-conjugated antibodies

3.1.1 Growth Conditions, Irradiation, and Harvesting of Cells

1. Culture the primary fibroblasts in complete culture medium supplemented with 10% FCS, 100 units/mL penicillin, and 100 μg/mL streptomycin at 37 °C in a humidified atmosphere with 5% CO_2. Grow the cells in 25 cm^2 flasks to approximately 70% confluence. For DSB induction, irradiate the cells at the dose of 5 Gy using X-ray-machine. Here we used RUM 17 X-ray machine operated at 200 kV and 12 mA with a 0.5 copper–aluminum filter.
2. Carefully remove medium from the flask and add a sufficient volume of sterile Versene solution to remove residual culture medium (*see* **Note 3**). Gently swirl the flask several times to rinse the cells, and then carefully aspirate Versene solution (*see* **Note 4**).
3. Using serological pipette, add minimal volume of 0.25% Trypsin-EDTA to cover the cells at the bottom of the flask. Roll flask gently to ensure that all cells are covered and place flask at 37 °C for approximately 2 min. Check cell detachment with an inverted microscope (*see* **Note 5**).
4. Add some culture medium to the flask to inactivate the trypsin and pipette cell suspension several times. Place cell suspension in 15 mL conical tube and centrifuge at 290 × *g* for 6 min at RT.

3.1.2 Cell Fixation and Immunostaining

1. Aspirate supernatant from 15 mL tube and add 300 μL of freshly prepared 1× Fix/Perm Buffer working solution to each tube. Vortex samples for approximately 3 s.
2. Place cell suspension in tubes designed for flow cytometry or in Eppendorf tubes, and incubate for 40 min on ice (*see* **Note 6**).
3. Add 300 μL of 1× Perm/Wash Buffer directly to the fixed and permeabilized cells. Pellet the cells by centrifugation at 290 × *g* for 6 min at +4 °C. All the following centrifugations should be performed in the same way.
4. Aspirate the supernatant. Add 600 μL of 1× Perm/Wash Buffer to the pelleted cells, resuspend the cells by vortexing, and pellet them again by centrifugation. Aspirate the supernatant.
5. Add 100 μL of primary mouse monoclonal anti-phospho-histone H2AX (Ser139) antibody diluted 1:250 in 1× Perm/Wash Buffer. Vortex the tube for 10 s and incubate in 37 °C water bath for 20 min (*see* **Note 7**).
6. Wash cells by adding 600 μL of 1× Perm/Wash Buffer to antibody solution followed by vortexing and centrifugation.
7. Wash cells as described in Subheading 3.1.2, **step 4**.
8. Add 100 μL of secondary goat anti-mouse IgG (H + L) Alexa Fluor 488-conjugated antibody diluted 1:400 in 1× Perm/Wash Buffer. Vortex the tube for 10 s and incubate in 37 °C water bath for 20 min (*see* **Notes 8** and **9**).

9. Wash cells as described in Subheading 3.1.2, **steps 6** and 7. Aspirate the supernatant.
10. Resuspend cell pellet in 300 μL of PBS and add DAPI to the final concentration of 0.5 μg/mL (*see* **Note 10**). Keep cells in the darkness for at least 20 min at RT.

3.1.3 Flow Cytometry Analysis

Using a flow cytometer, analyze the cells and acquire data on fluorescence intensity and cell cycle distribution (*see* **Note 11**).

For data analysis, a minimum of 10,000–15,000 events are required.

Analyze samples of control and irradiated cells sequentially.

Alexa Fluor 488 and DAPI are excited by different lasers, their spectra do not overlap, and, therefore, compensation is not necessary.

1. Use forward-scattered light (forward scatter or FSC) and side-scattered light (side scatter or SSC) to choose a cell population of interest. Determine the population of cells by analysis of cell distribution on the scatter plot of FSC versus SSC. Gate undamaged cells and exclude cellular debris and cell aggregates located above and below the gated area (Gt1) of the dot plot. The cellular debris has lower FSC and SSC than undamaged cells, and cell clamps and/or doublets have higher FSC and SSC than single cells (*see* Fig. 1a(i), b(i)).
2. The main population of cells gated on the scatter plot of FSC versus SSC may contain not only single cells, but also a number of cell aggregates. On the dot plot of 405 nm laser area (DAPI-A) versus 405 nm laser height (DAPI-H) (not shown), analyze the cell population to distinguish single cells, doublets, triplets, and so on. Select single cells by gating.
3. Using control cells, obtain univariate histogram plot of 488 nm laser area (Alexa Fluor 488-A) in logarithmic scale versus cell count (not shown) to determine background signal and exclude autofluorescence of cells. Gate positively stained cells.
4. Analyze gated cells on histogram plot DAPI-A versus cell count to see distribution of cells in the cell cycle (*see* Fig. 1a(ii), b(ii)).
5. Using control cells, analyze gated cells on bivariate plot DAPI-A versus Alexa Fluor 488-A, gate stained and unstained cells and estimate their percentage. Negative cells should represent around 94–98% of the cell population (Gt3), while positively stained cells (Gt4)—around 2–6% (*see* Fig. 1a(iii)).
6. Repeat **steps 1–5** for irradiated cells. For irradiated cells, on bivariate histogram plot DAPI-A versus Alexa Fluor 488-A, use the same level of gating as for the control cells (*see* Fig. 1b(iii), Gt5, Gt6).
7. Estimate mean fluorescence intensity per cell for G1 and G2 cell populations.

3.2 Protocol 2. Detection of γ-H2AX Using Saponin-Based Permeabilization

In this protocol, γ-H2AX staining is performed after cell fixation with formaldehyde and subsequent permeabilization with saponin solution. Permeabilization conditions described here do not change the ratio between positive and negative populations of irradiated cells compared to data obtained using Protocol 1. The procedure takes around 2 h 15 min excluding the time needed for harvesting the cells.

1. Prepare the cells for fixation as described in Subheading 3.1.1, **steps 2, 3** and **4**. It is optimal to start with 10^6 cells per sample.
2. Aspirate supernatant from 15 mL conical tubes, add 600 μL of PBS, vortex cells, and transfer the suspension to an Eppendorf tube.
3. Pellet the cells by centrifugation at 290 × *g* for 6 min at +4 °C. All subsequent centrifugations are performed in a similar way.
4. Add 600 μL of freshly prepared 4% formaldehyde in PBS precooled to +4 °C. Vortex briefly and incubate for 10 min at +4 °C.
5. Add 300 μL of cold (precooled to +4 °C) PBS to cell suspension in formaldehyde solution, vortex and pellet cells by centrifugation. Aspirate the supernatant.
6. Wash cells by adding 600 μL of cold PBS to cell pellet, followed by vortexing and centrifugation. After centrifugation, aspirate the supernatant.
7. Vortex the pellet and add 750 μL of permeabilization solution (Perm) containing 0.1% saponin and 1–10% FCS in PBS. Incubate for 30 min at +4 °C. Tap the tube gently with your finger several times during incubation period (*see* **Notes 12** and **13**).
8. Pellet cells by centrifugation, aspirate the supernatant, and add 100 μL of primary mouse monoclonal anti-phospho-histone H2AX (Ser139) antibody diluted 1:250 in Perm. Vortex for 3 s and incubate for 20 min at 37 °C in the water bath. Tap the tube gently with your finger a couple of times during incubation (*see* **Note 14**).
9. Add 600 μL of Perm to cell suspension in antibody solution, vortex and centrifuge the cells. Aspirate the supernatant.
10. Wash cells by adding 600 μL of Perm to the cell pellet followed by vortexing and centrifugation. Aspirate the supernatant.
11. Add 100 μL of secondary goat anti-mouse IgG (H + L) Alexa Fluor 488-conjugated antibody diluted 1:400 in Perm. Vortex for 3 s and incubate for 20 min at 37 °C in the water bath.
12. Repeat **steps 9** and **10** of this section.
13. Add 300 μL of cold PBS (*see* **Note 15**).

14. For DNA counterstaining, add 3 μL of 10% Triton X-100 and 3 μL of 100 μg/mL DAPI stock solution (final concentrations 0.1% Triton X-100 and 0.5 μg/mL DAPI). Vortex and keep in the darkness for 20 min at RT before analysis (*see* **Note 16**).
15. Perform flow cytometry analysis as described in Subheading 3.1.3.

4 Notes

1. After trypsinization, avoid Ca^{2+} and Mg^{2+} in PBS, because they promote cell adhesion. You can use any commercial PBS or prepare it yourself from separate compounds. Formulation for 1 L of PBS (10×) is: 80 g NaCl, 2 g KCl, 21.6 g $Na_2HPO_4 \times 7H_2O$, 2 g KH_2PO_4. When diluted to 1× PBS with H_2O, its pH is 7.4.
2. We routinely use BD Pharmingen Transcription Factor Buffer Set in our laboratory, because it is optimal for flow cytometry analysis of different intracytoplasmic and intranuclear proteins.
3. In our experience, an initial number of cells of 5×10^5 cells per sample is enough for flow cytometry analysis of γ-H2AX fluorescence intensity.
4. Versene is an EDTA solution which removes Ca^{2+} ions from the cell surface, thus improving detachment of cells. Detachment of primary fibroblasts from the dish surface is faster when Versene solution is used instead of PBS.
5. Over-trypsinization can severely damage cells. Thus, it is essential to check cells for detachment progress under the microscope.
6. In case of low number of cells, for easier visualization of cell pellets from small amounts of cells, it is more convenient to place cells in 1.5 mL Eppendorf tubes instead of 5 mL round-bottom tubes designed for flow cytometry.
7. Some researchers noted that antibody concentration for flow cytometry was usually higher than that for immunofluorescence [7]. In our protocol, the dilution of primary and secondary antibodies was the same as was used for γ-H2AX detection by immunofluorescence microscopy. It is most likely that the degree of dilution depends on several parameters: quality of antibodies, fixation/permeabilization conditions, and the type of cell culture.
8. Staining with secondary antibodies (Subheading 3.1.2, **steps 8** and **9**) is omitted when Alexa Fluor 488-conjugated primary antibodies to γ-H2AX are used. It should be noted, however, that the use of mouse monoclonal anti-phospho-histone H2AX

(Ser139) Alexa Fluor 488-conjugated antibody, cat # 05-636AF488 (Millipore) resulted in the decrease of positively stained cell population analyzed after irradiation in comparison with indirect staining described in our protocol.

9. In our modified protocol, we used 20 min incubation at 37 °C with primary and secondary antibodies instead of 40 min incubation at +4 °C suggested by BD Pharmingen protocol. Flow cytometry analysis has shown that the results obtained by both methods do not differ.
10. Cells resuspended in PBS can be stored for several days at +4 °C in the darkness before analysis.
11. This protocol is optimized for mouse monoclonal anti-phospho-histone H2AX (Ser139) antibody, cat # 05-636 (Millipore). If other commercial antibodies are used, optimize the degree of their dilution for different cell lines.
12. Addition of FCS is critical and is needed to prevent cell loss during incubations, especially for incubations at 37 °C.
13. Saponin concentration has to be optimized for different cell lines. We found that the optimal final concentration of saponin is 0.1% for human fibroblasts. The percentage of stained cells is decreased if you use 0.05% saponin for permeabilization. The increase of saponin concentration to 0.2% leads to progressive cell loss during subsequent centrifugations.
14. Saponin treatment leads to the formation of permanent and transient cell membrane openings [4]. The latter ones disappear during subsequent incubations of cells in the absence of saponin. Therefore, saponin is present in solutions during all steps of cell preparation for flow cytometry analysis including the treatment with antibody.
15. It is possible to store fixed and stained cells in PBS during several days at +4 °C before DNA counterstaining with DAPI and flow cytometry analysis.
16. The cells can be analyzed for flow cytometry directly in Perm buffer. In this case, G1 and G2 cell populations cannot be distinguished after DAPI counterstaining. Therefore, additional permeabilization with Triton X-100 is needed for visualization of cell cycle distribution.

Acknowledgments

This work was supported by the Russian Science Foundation (Grant # 16-14-10240).

References

1. Rogakou EP, Pilch DR, Orr AH et al (1998) DNA double-stranded breaks induce histone H2AX phosphorylation on serine 139. J Biol Chem 273:5858–5868
2. Rothkamm K, Löbrich M (2003) Evidence for a lack of DNA double-strand break repair in human cells exposed to very low x-ray doses. Proc Natl Acad Sci U S A 100:5057–5062
3. http://www.bdbiosciences.com/ds/pm/tds/562725.pdf
4. Krutzik PO, Irish JM, Nolan GP et al (2004) Analysis of protein phosphorylation and cellular signaling events by flow cytometry: techniques and clinical applications. Clin Immunol 110:206–221
5. Melan MA (1999) Overview of cell fixatives and cell membrane permeants. Methods Mol Biol 115:45–55
6. Muslimovic A, Ismail IH, Gao Y et al (2008) An optimized method for measurement of gamma-H2AX in blood mononuclear and cultured cells. Nat Protoc 3:1187–1193
7. Forment JV, Jackson SP (2015) A flow cytometry-based method to simplify the analysis and quantification of protein association to chromatin in mammalian cells. Nat Protoc 10:1297–1307

Chapter 12

Rapid Detection of DNA Strand Breaks in Apoptotic Cells by Flow- and Image-Cytometry

Hong Zhao and Zbigniew Darzynkiewicz

Abstract

Extensive DNA fragmentation that generates a multitude of DNA double-stand breaks (DSBs) is a hallmark of apoptosis. We developed several variants of the widely used TUNEL methodology that is based on the use of exogenous terminal deoxynucleotidyl transferase (TdT) to label 3′OH ends in DSBs with fluorochromes. Flow- or image-cytometry is then employed to detect and quantify apoptotic cells labeled this way. Here, we describe a variant of this technique using BrdUTP as a TdT substrate. The incorporated BrdU is subsequently visualized by a fluorochrome-tagged antibody. This is a particularly simple, rapid, and sensitive approach to detect DSBs.

We also describe modifications of the labeling protocol permitting the use of deoxyribonucleotides other than BrdUTP to label DSBs. Concurrent differential staining of cellular DNA and multiparameter analysis of cells by flow- or image-cytometry enable correlations between apoptosis induction and the cell cycle phase. Examples of the detection of apoptotic cells in cultures of human leukemic cell lines treated with TNF-α and DNA topoisomerase I inhibitor topotecan are presented. The protocol can be applied to cells treated with cytotoxic drugs in vitro, ex vivo, or to clinical samples.

Key words Apoptosis, DNA damage, Flow cytometry, Laser scanning cytometry, Cell cycle, Immunofluorescence, BrdU

1 Introduction

During apoptosis DNA undergoes extensive fragmentation at internucleosomal linker regions which generates a multitude of DNA double-strand breaks (DSBs) [1, 2]. Their presence is considered to be one of the most characteristic markers of apoptotic cells. A widely used approach to identify apoptotic cells, thus, relies on labeling DSBs in situ either with fluorochromes [3–5] or absorption dyes [6]. We have developed several variants of the methodology that is based on the use of exogenous terminal deoxynucleotidyl transferase (TdT) to label 3′OH termini of the DSBs either indirectly or directly with fluorochrome-tagged deoxyribonucleotides, commonly defined as the TUNEL assay [2–8]. In this Chapter, we describe the variant based on DSBs labeling with

Vladimir V. Didenko (ed.), *Fast Detection of DNA Damage: Methods and Protocols*, Methods in Molecular Biology, vol. 1644, DOI 10.1007/978-1-4939-7187-9_12,

BrdUTP that subsequently is detected immunocytochemically with BrdU antibody (Ab). The BrdUTP labeling assay offers much greater sensitivity than other TUNEL variants [9]. However, modifications of the protocol that allow use of deoxyribonucleotides other than BrdUTP also are described. Concurrent staining of cellular DNA with propidium iodide (PI) or 4′,6-diamidino-2-phenylindole (DAPI) and multiparameter analysis of cells by flow- or image-cytometry enables correlation of the cell cycle phase with the induction of apoptosis [10]. The protocol can be applied to cells growing in vitro, treated ex vivo with cytotoxic drugs as well as to clinical samples (*see* **Note 1**). The method presented in this Chapter can be applied to suspended cells and to cells attached to microscope slides. The suspended cells are measured by flow cytometry, whereas the attached cells are analyzed by image cytometry, e.g., using an instrument such as the laser scanning cytometer (LSC). LSC is a microscope-based cytofluorometer capable of rapid, sensitive, and accurate measurement of individual cell fluorescence [11]. Cells staining on slides eliminates their loss that otherwise occurs during repeated centrifugations in sample preparation for flow cytometry. Therefore, the procedure offers an advantage when applied to samples with paucity of cells such as fine needle aspirate or spinal fluid tap (*see* **Note 2**). Another advantage of LSC it that it offers a possibility of electronic selection (gating) of cells of interest during the initial measurement for their subsequent analysis by imaging or staining with other fluorochromes. Imaging and visual examination are of particular importance because the characteristic changes in cell morphology are considered the gold standard for positive identification of apoptotic cells [3, 4]. Furthermore, the cell attributes measured by LSC on live cells can be correlated with the attributes that generally require cell fixation to be measured [12]. For example, activation of caspases [12–14], DNA replication [15], translocation of Bax to mitochondria [16], or activation of NF-κB transcription factor [17], the key events associated with apoptosis, can be correlated, in the very same cells, with the presence of apoptosis-associated DSBs as well as with the position of cells in the cell cycle.

Fixation and permeabilization of the cells are the initial essential steps required to successfully label DSBs. Cells are briefly fixed with the crosslinking fixative formaldehyde and then permeabilized by suspending in ethanol or using detergents in the subsequent rinses. By crosslinking small DNA fragments to other cell constituents, formaldehyde prevents their extraction, which otherwise occurs during repeated centrifugations and rinses [18]. The 3′OH termini of the DSBs serve as primers and become labeled in this procedure with BrdU when incubated with BrdUTP in the reaction catalyzed by exogenous TdT [2]. The incorporated BrdU is immunocytochemically detected by BrdU Ab conjugated to fluorochromes [9]. The BrdU Ab is a widely available reagent, also used in studies of cell proliferation to detect BrdU incorporated during DNA replication [19, 20]. The sensitivity of DSBs detection is higher and the overall cost of reagents is significantly lower when BrdUTP is used, as compared to the alternative labeling with biotin- (or digoxigenin-) [3, 4] or by directly fluorochrome-tagged deoxyribonucleotides [8]. Furthermore, the time of incubation with anti BrdU antibodies can be shortened to 20–30 min, if needed, due to the high sensitivity of the antibody-based detection of the BrdU epitope.

2 Materials

2.1 Reagents and Glassware

1. Phosphate-buffered saline (PBS), pH 7.4.
2. 1% Formaldehyde (methanol-free, "ultrapure"), in PBS, pH 7.4.
3. 70% Ethanol.
4. TdT (Roche Diagnostics). Supplied in storage buffer: 60 mM potassium phosphate at pH 7.2, 150 mM KCl, 1 mM 2-mercaptoethanol and 0.5% Triton X-100, 50% glycerol). The 5× TdT reaction buffer contains: 1 M potassium (or sodium cacodylate) 125 mM HCl, pH 6.6 (Roche Diagnostics), and 1.25 mg/mL bovine serum albumin (BSA).
5. 5-Bromo-2′-deoxyuridine-5′-triphosphate (BrdUTP) stock solution (50 μL): 2 mM BrdUTP (Sigma) in 50 mM Tris–HCl, pH 7.5.
6. 10 mM $CoCl_2$ (Sigma).
7. Rinsing buffer: 0.1% Triton X-100 (Sigma) and 5 mg/mL BSA dissolved in PBS.
8. Alexa Fluor 488-conjugated anti-BrdU monoclonal antibody (mAb): Dissolve 1.0 μg of Alexa Fluor 488-conjugated anti-BrdU Ab in 100 μL of PBS containing 0.3% Triton X-100 and 1% (w/v) BSA. Alternatively, use Fluorescein- (FITC)- or Alexa Fluor 647-conjugated anti-BrdU Ab (Phoenix Flow System (San Diego, CA), or Molecular Probes/ThermoFisher).
9. PI staining buffer: 5 μg/mL PI (Molecular Probes/ThermoFisher), 10 μg/mL of RNase A (DNase-free) (Sigma) in PBS. Alternatively, use 1 μg/mL solution of 4′,6-diamidino-2-phenylindole (DAPI) in PBS.
10. Microscope slides or single- or multi-chambered Falcon CultureSlides (BD Biosciences) (to be used in conjunction with analysis by LSC/iCys).
11. Coplin jars (to be used in conjunction with analysis by LSC/iCys).
12. Parafilm "M" (to be used in conjunction with LSC/iCys).
13. Glycerol (to be used in conjunction with analysis by LSC/iCys).

14. Single- or multi-chambered Falcon Culture Slides (BD Biosciences).
15. BODIPY-FL-X-dUTP, fluorescein-, Cascade Blue-, Texas Red-dUTPs or other ChromaTide nucleotides (Thermo Fisher) or CY-3-dCTP or other cyanine dye conjugates.

2.2 Commercial Kits

Several kits for labeling DSBs are commercially available. The APO-BRDU kit (Phoenix Flow Systems, San Diego, CA) uses a BrdUTP methodology similar to that described in this chapter. As mentioned, this methodology offers the most sensitive means of DNA strand break detection [9]. The APO-DIRECT kit (also from Phoenix) offers a single-step labeling of DNA strand breaks with the fluorochrome-tagged deoxynucleotide. Its virtue is simplicity and shortened overall incubation time, but it is less sensitive than the APO-BRDU. Of particular importance is that the positive and negative control cells are supplied with each of these Phoenix kits. It should be noted that kits developed and distributed by Phoenix Flow Systems are also provided by other vendors. The kits utilizing biotin- or digoxigenin-tagged dUTP are also commercially available.

2.3 Instrumentation

1. Flow cytometers of different types, offered by several manufacturers, can be used to measure cell fluorescence following staining according to the procedures described below. The manufacturers of the most common flow cytometers are Coulter/Beckman Corporation (Miami, FL), BD Biosciences (formerly Becton Dickinson Immunocytometry Systems; San Jose, CA), iCyt (Urbana-Champain, IL), and PARTEC GmbH (Zurich, Switzerland). The multiparameter Laser Scanning Cytometer (LSC; iCys™ model) is available from Thorlabs, Inc., (Newton, NJ, USA). Cytospin centrifuge, which is used in conjunction with LSC/iCys, is provided by Shandon (Pittsburgh, PA).
2. The software to deconvolve the DNA content frequency histograms, to analyze the cell cycle distributions, is available from Phoenix Flow Systems or Verity Software House (Topham, MA).

3 Methods

3.1 DNA Strand Break Labeling with BrdUTP for Analysis by Flow Cytometry

1. Suspend 1–2 × 10^6 cells in 0.5 mL PBS. With a Pasteur pipette transfer this suspension into a 5 mL polypropylene tube (*see* **Note 2**) containing 4.5 mL of ice-cold 1% formaldehyde (*see* **Note 3**). Keep the tube for 15 min on ice.
2. Centrifuge at 300 × *g* for 5 min and resuspend cell pellet in 5 mL of PBS. Centrifuge again and resuspend cell pellet in 0.5 mL of PBS. With a Pasteur pipette transfer the suspension to a tube containing 4.5 mL of ice-cold 70% ethanol. The cells can be stored in ethanol, at −20 °C for several weeks.

3. Centrifuge at 200 × *g* for 3 min, remove ethanol, resuspend cells in 5 mL of PBS, and centrifuge at 300 × *g* for 5 min.
4. Resuspend the pellet in 50 μL of a solution containing:
   - 10 μL TdT 5× reaction buffer.
   - 2.0 μL of BrdUTP stock solution.
   - 0.5 μL (12.5 units) TdT.
   - 5 μL $CoCl_2$ solution.
   - 33.5 μL distilled H_2O.
5. Incubate the cells in this solution for 40 min at 37 °C (*see* **Notes 4** and **5**).
6. Add 1.5 mL of the rinsing buffer, and centrifuge at 300 × *g* for 5 min.
7. Resuspend cell pellet in 100 μL of Alexa Fluor 488-conjugated anti-BrdU Ab solution. [Alternatively, you may use the Ab conjugated either with fluorescein (FITC) or Alexa Fluor 647].
8. Incubate at room temperature for 1 h.
9. Add 1 mL of PI staining solution (alternatively you may add 1 mL of the DAPI staining solution).
10. Incubate for 30 min at room temperature, or 20 min at 37 °C, in the dark.
11. Analyze cells by flow cytometry.
    - Illuminate with blue light (488 nm laser line or BG12 excitation filter).
    - Measure green fluorescence of FITC- (or Alexa Fluor 488)-conjugated anti BrdU Ab at 530 ± 20 nm.
    - Measure intensity of red fluorescence of PI at >600 nm. Alternatively, if DNA was stained with DAPI instead of PI use UV light or near UV light diode as an excitation source and measure intensity of blue fluorescence (480 ± 20 nm).

The bi-variate (DSBs versus cellular DNA content) distributions (scatterplots) illustrating the cell populations containing a fraction of apoptotic cells labeled according to the method described in the protocol and analyzed by flow cytometry are shown in Fig. 1, and analyzed by LSC (iCys) as shown in Fig. 2.

The bivariate (DSBs versus cellular DNA content) distributions (scatterplots) illustrating the cell populations containing a fraction of apoptotic cells labeled according to the method described in the protocol and analyzed by flow cytometry are shown in Fig. 1, and analyzed by LSC (iCys) are shown in Fig. 2. A correlation between the induction of apoptosis and cell position in the cell cycle is clearly evident: in the case of topotecan-treated HL-60 cells nearly all apoptotic cells are S-phase cells (Fig. 1) while the apoptotic U-932 cells treated with TNF-α are predominantly G_1- and early-S phase cells (Fig. 2).

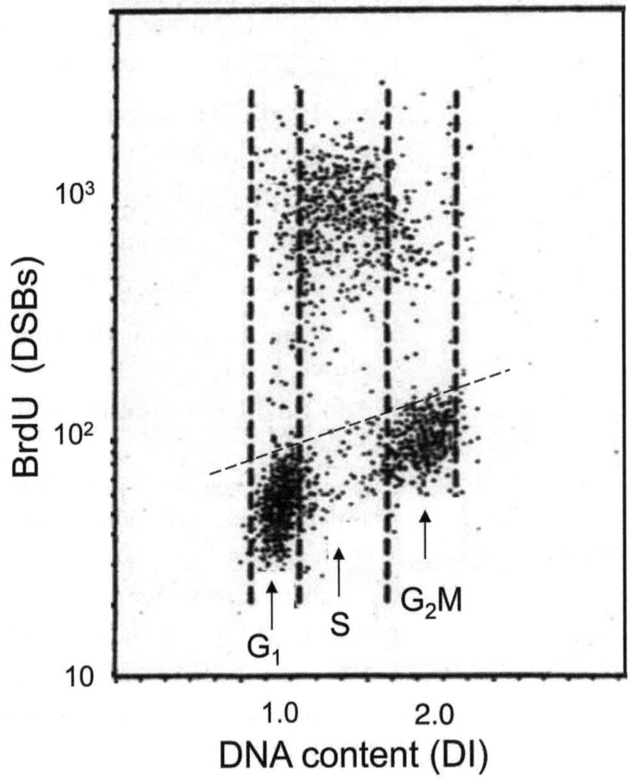

Fig. 1 Detection of apoptotic cells after DSBs labeling with BrdUTP and fluorescence analysis by flow cytometry. To induce apoptosis leukemic HL-60 cells were treated in culture with DNA topoisomerase I inhibitor topotecan (0.15 μM) for 4 h. The cells were then subjected to DSBs labeling with BrdUTP as described in the protocol using fluorescein-tagged BrdU Ab and staining DNA with PI. Cellular fluorescence was measured by flow cytometry. The data are presented as the bivariate distributions (scatterplots) illustrating cellular DNA content (DNA index, DI) versus DSBs labeled with BrdU Ab. Note that essentially only S-phase cells underwent apoptosis as shown by high intensity of their BrdU-associated fluorescence, above the control level marked by the *skewed dashed line*. The leukemic cells treated with topoisomerase I inhibitors topotecan or camptothecin for 3–5 h and processed as described in this protocol present a convenient experimental model to assess whether the DSBs labeling is effective because in the same cell population there are DSBs positive (S-phase) and negative (G_1 and G_2M) cells

3.2 DSBs Labeling with Other Markers for Analysis by Flow Cytometry

As mentioned in Subheading 1, DNA strand breaks can be labeled with deoxynucleotides tagged with a variety of other fluorochromes. Several types of dUTP conjugates, including BODIPY dyes (e.g., BODIPY-FL-X-dUTP), fluorescein, Cascade Blue and Texas Red and cyanine dyes conjugates (e.g., CY-3-dCTP), are commercially available (*see* Subheading 2.1, **item 15**). Indirect labeling, via biotinylated- or digoxigenin-conjugated deoxyribonucleotides, is also possible by using commercially available fluorochrome-conjugated

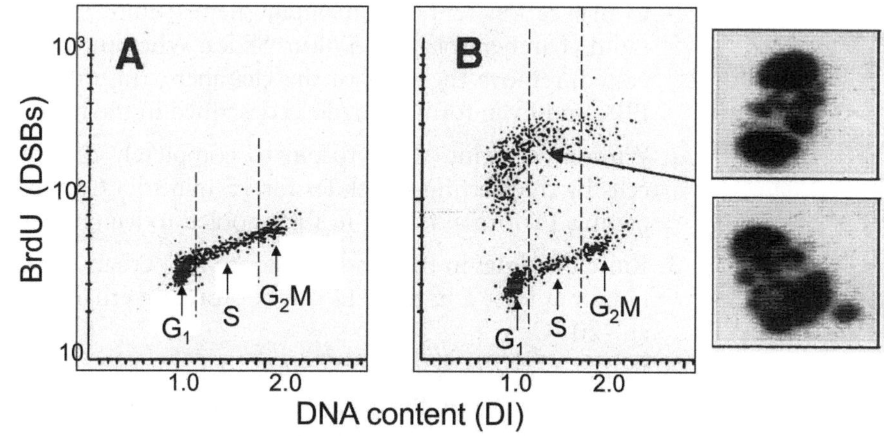

Fig. 2 Detection of apoptotic cells after DSBs labeling with BrdUTP and analysis by LSC. U-937 cells were untreated (**a**) or treated with tumor necrosis factor-α (TNF-α) in the presence of cycloheximide (**b**, refs. 24, 25). The cells were then subjected to DNA strand break labeling and DNA staining as described in the protocol using fluorescein-tagged anti-BrdU Ab and staining of DNA with PI in the presence of RNase. Cell fluorescence was measured by LSC. The bivariate distributions (scatterplots) identify apoptotic cells as the cells with labeled DSBs (strong green fluorescence intensity), and also reveal the cell cycle position of cells in either apoptotic or nonapoptotic population. Note predominance of G_1 and S phase cells among apoptotic cells. The cells with strong DSBs labeling were relocated, imaged by LSC, and their representative images are shown. These cells show nuclear fragmentation and chromatin condensation, the typical features of apoptosis [3, 4]

avidin, or streptavidin, as well as anti-digoxigenin antibodies. DNA strand breaks, thus, can be labeled with a dye of any desired fluorescence emission and excitation wavelength.

The procedure described in Subheading 3.1 can be adopted to utilize any of these fluorochromes.

In the case of the direct labeling [8], the fluorochrome-conjugated deoxyribonucleotide is included in the reaction solution (0.25–0.5 nmol per 50 μL) instead of BrdUTP, as described in **step 4** of Subheading 3.1. Following the incubation step (**step 5**), omit **steps 6–8**, and stain cells directly with PI (**step 9**).

In the case of the indirect labeling, instead of BrdUTP, digoxigenin- or biotin-conjugated deoxyribonucleotides are included into the reaction buffer (0.25–0.5 nmol per 50 μL) at **step 4**. The cells are then incubated either with the fluorochrome-conjugated anti-digoxigenin mAb (0.2–0.5 μg per 100 μL of PBS containing 0.1% Triton X-100 and 1% BSA), or with fluorochrome-conjugated avidin or streptavidin (0.2–0.5 μg per 100 μL, as above) at **step 7** and then processed through **steps 8–10** as described in the protocol. Analysis by flow cytometry is carried out with excitation and emission wavelengths appropriate for the used fluorochrome.

3.3 DNA Strand Break Labeling for Analysis by LSC (iCys®)

1. Transfer 300 μL of cell suspension (in tissue culture medium, with serum) containing approximately 20,000 cells into a cytospin chamber. Cytocentrifuge at 1000 rpm (~150 × *g*) for 6 min to deposit the cells on a microscope slide. (Alternatively,

to analyze adherent cells, maintain them in cultures in single- or multi-chambered Falcon CultureSlides. When the cells are harvested, remove the walls of the chambers, rinse the cells with PBS, and fix in formaldehyde as described in the next step.)

2. Without allowing the cytospin to completely dry, prefix the cells by transferring the slides for 15 min to a Coplin jar containing 1% formaldehyde in PBS, cooled to ice temperature.
3. Rinse the slides in PBS and transfer to 70% ethanol; fix in ethanol for at least 1 h; the cells can be stored in ethanol for weeks at −20 °C.
4. Follow **steps 4–8** of Subheading 3.1 as described for flow cytometry. Carefully layer small volumes (approximately 100 μL) of the respective buffers, rinses, or staining solutions onto the cytospin area of the horizontally placed slides or over the sites of individual chambers if the cells were grown on Chamber Slides. At appropriate times remove these solutions with Pasteur pipette (or vacuum suction pipette). To prevent drying, place a 2 × 4 cm strip of Parafilm over the site where the cells are present atop of the solutions used for cell incubations (*see* **Note 6**).
5. Replace the PI staining solution with a drop of a mixture of glycerol and PI staining solution (9:1) and mount under the coverslip. To preserve the specimen for longer period of time or transport, seal the coverslip with nail polish or melted paraffin.
6. Measure cell fluorescence on LSC.
   - Excite fluorescence with 488 nm laser line.
   - Measure green fluorescence of Alexa Fluor 488 or fluorescein-anti BrdU Ab at 530 ± 20 nm.
   - Measure red fluorescence of PI at >600 nm.

(Alternatively, if DSBs are labeled with Alexa Fluor 647 excite fluorescence with red diode laser and measure fluorescence intensity at far-red wavelength. If DAPI is used to stain DNA, excite DAPI fluorescence with UV or near UV light laser and measure fluorescence intensity at 480 ± 20 nm wavelength).

The typical results are shown in Figs. 1 and 2 (*see* **Notes** 7 and **8** and 7).

3.4 Controls

The procedure for labeling DNA strand beaks involves many reagents. Negative results, therefore, may not necessarily mean the absence of DNA strand breaks (*see* **Note** 7) but may be due to methodological problems, such as the loss of TdT activity, degradation of BrdUTP, etc. It is necessary, therefore, to include a positive and negative control. An excellent control is to use HL-60 cells treated (during their exponential growth) for 3–4 h with 0.2 μM of the DNA topoisomerase I inhibitor camptothecin (CPT) or topotecan (TPT). Because under these conditions only DNA replicating (S-phase) cells undergo apoptosis the cells in G_1 and G_2/M may serve as negative control populations, whereas the S phase cells in the same sample represent the positive control. As mentioned in Subheading 2.2, the APO-DIRECT kit provides already fixed cells pretreated with CPT that contain positive (S-phase) and negative (G_1 and G_2M) cells in a single sample.

Another negative control contains cells processed identically as described in Subheading 3.1 except that TdT is excluded from **step 4**.

4 Notes

1. This method is also useful for clinical material, such as samples obtained from leukemias, lymphomas, and solid tumors [21, 22], and can be combined with surface immunophenotyping. When the methods are combined the cells are first immunophenotyped, then fixed with 1% formaldehyde (which stabilizes the antibody bound on the cell surface) and subsequently subjected to the DSBs detection assay using other color fluorochrome (*see* Subheading 3.1) than the one used for immunophenotyping. The percent of apoptotic (DSBs-positive) cells is then estimated within the gated-immunophenotype cell population.
2. If the sample initially contains a small number of cells, cell loss during repeated centrifugations can often be a problem. We recommend using polypropylene, or siliconized glass tubes to minimize cell loss. Since transferring cells from one tube to another results in electrostatic attachment of a large fraction of cells to the surface of each new tube all steps of the procedure (including fixation) should be done in the same tube. Addition of 1% (w/v) BSA into rinsing solutions also decreases cell loss. When the sample contains very few cells, the carrier cells, which later can be recognized based on differences in DNA content (e.g., chicken erythrocytes), may be included. Because there is no cell loss during processing for analysis by LSC the samples with paucity of cells can easily be measured.
3. Cell pre-fixation with a crosslinking agent such as formaldehyde is required to prevent the loss of the fragmented DNA from apoptotic cells [2]. This ensures that despite the repeated cell washings during the procedure, the DNA content of apoptotic cells (and with it the number of DSBs) is not markedly diminished.
4. Alternatively, incubate at room temperature overnight.
5. Control cells should be incubated in the same solution, but without TdT.

6. It is generally easy to identify apoptotic cells, due to their intense labeling with Alexa Fluor 488, fluorescein, or Alexa Fluor 647 conjugated anti-BrdU Ab. The high fluorescence intensity often requires use of the exponential scale (logarithmic amplifiers of the flow cytometer or LSC) for data acquisition and display (Figs. 1 and 2). The figures show that after the cellular DNA content of each, apoptotic and nonapoptotic cell population is measured, the cell cycle distribution and/or DNA ploidy of these both populations can be estimated.
7. While the presence of extensive DNA breakage is marked by strong fluorescence after TUNEL labeling, and is a very characteristic feature of apoptosis, weak fluorescence may not necessarily mean the lack of apoptosis. In some cell systems DNA fragmentation stops at 300–50 kb size DNA fragments and does not progress into the internucleosomal linker regions [23].
8. It is essential that the incubations are carried out in a humidified chamber. Even minor drying produces severe artifacts.

Acknowledgment

Supported by NCI grant RO1 28 704 and the Robert A. Welke Cancer Research Foundation.

References

1. Nagata S (2000) Apoptotic DNA fragmentation. Exp Cell Res 256:12–18
2. Kajstura M, Halicka HD, Pryjma J, Darzynkiewicz Z (2007) Discontinuous fragmentation of nuclear DNA during apoptosis revealed by discrete "sub-G1" peaks on DNA content histograms. Cytometry A 71A:125–131
3. Gorczyca W, Bruno S, Darzynkiewicz RJ, Gong J, Darzynkiewicz Z (1992) DNA strand breaks occurring during apoptosis: Their early *in situ* detection by the terminal deoxynucleotidyl transferase and nick translation assays and prevention by serine protease inhibitors. Int J Oncol 1:639–648
4. Gorczyca W, Gong JP, Darzynkiewicz Z (1993) Detection of DNA strand breaks in individual apoptotic cells by the in situ terminal deoxynucleotidyl transferase and nick translation assays. Cancer Res 53:1945–1951
5. Darzynkiewicz Z, Juan G, Li X, Gorczyca W, Murakami T, Traganos F (1997) Cytometry in cell necrobiology: analysis of apoptosis and accidental cell death (necrosis). Cytometry 27:1–20
6. Wlodkowic D, Telford W, Skommer J, Darzynkiewicz Z (2011) Apoptosis and beyond: cytometry in studies of programmed cell death. Methods Cell Biol 103:55–98
7. Darzynkiewicz Z, Galkowski D, Zhao H (2008) Analysis of apoptosis by cytometry using TUNEL assay. Methods 44:250–254
8. Li X, Traganos F, Melamed MR, Darzynkiewicz Z (1995) Single step procedure for DNA strand break labeling. Detection of apoptosis and DNA replication. Cytometry 20:172–180
9. Li X, Darzynkiewicz Z (1995) Labelling DNA strand breaks with BrdUTP. Detection of apoptosis and cell proliferation. Cell Prolif 28:571–579
10. Gorczyca W, Gong J, Ardelt B, Traganos F, Darzynkiewicz Z (1993) The cell cycle related differences in susceptibility of HL-60 cells to apoptosis induced by various antitumor drugs. Cancer Res 53:3186–3192
11. Pozarowski P, Holden E, Darzynkiewicz Z (2013) Laser scanning cytometry: principles and applications. An update. Methods Mol Biol 913:187–212

12. Li X, Darzynkiewicz Z (1999) The Schrödinger's cat quandary in cell biology: integration of live cell functional assays with measurements of fixed cells in analysis of apoptosis. Exp Cell Res 249:404–412
13. Li X, Du L, Darzynkiewicz Z (2000) Caspases are activated during apoptosis independently of dissipation of mitochondrial electrochemical potential. Exp Cell Res 257:290–297
14. Huang X, Okafuji M, Traganos F, Luther E, Holden E, Darzynkiewicz Z (2004) Assessment of histone H2AX phosphorylation induced by DNA topoisomerase I and II inhibitors topotecan and mitoxantrone and by DNA crosslinking agent cisplatin. Cytometry A 58A:99–110
15. Li X, Melamed MR, Darzynkiewicz Z (1996) Detection of apoptosis and DNA replication by differential labeling of DNA strand breaks with fluorochromes of different color. Exp Cell Res 222:28–37
16. Bedner E, Li X, Kunicki J, Darzynkiewicz Z (2000) Translocation of Bax to mitochondria during apoptosis measured by laser scanning cytometry. Cytometry 41:83–88
17. Deptala A, Bedner E, Gorczyca W, Darzynkiewicz Z (1998) Simple assay of activation of nuclear factor kappa B (NF-κB) by laser scanning cytometry (LSC). Cytometry 33:376–382
18. Gong J, Traganos F, Darzynkiewicz Z (1994) A selective procedure for DNA extraction from apoptotic cells applicable for gel electrophoresis and flow cytometry. Anal Biochem 218:314–319
19. Dolbeare F, Selden JR (1994) Immunochemical quantitation of bromodeoxyuridine: application to cell cycle kinetics. Methods Cell Biol 41:297–316
20. Darzynkiewicz Z, Traganos F, Zhao H, Halicka HD, Li J (2011) Cytometry of DNA replication and RNA synthesis: historical perspective and recent advances based on "click chemistry". Cytometry A 79A:328–337
21. Gorczyca W, Bigman K, Mittelman A, Ahmed T, Gong J, Melamed MR, Darzynkiewicz Z (1993) Induction of DNA strand breaks associated with apoptosis during treatment of leukemias. Leukemia 7:659–670
22. Li X, Gong J, Feldman E, Seiter K, Traganos F, Darzynkiewicz Z (1994) Apoptotic cell death during treatment of leukemias. Leuk Lymphoma 13:65–72
23. Oberhammer F, Wilson JW, Dive C, Morris ID, Hickman JA, Wakeling AE, Walker PR, Sikorska M (1993) Apoptotic death in epithelial cells: cleavage of DNA to 300 and 50 kb fragments prior to or in the absence of internucleosomal degradation of DNA. EMBO J 12:3679–3684
24. Li X, Darzynkiewicz Z (2000) Cleavage of poly(ADP-ribose) polymerase measured *in situ* in individual cells: relationship to DNA fragmentation and cell cycle position during apoptosis. Exp Cell Res 255:125–132
25. Bedner E, Smolewski P, Amstad P, Darzynkiewicz Z (2000) Activation of caspases measured in situ by binding of fluorochrome-labeled inhibitors of caspases (FLICA): correlation with DNA fragmentation. Exp Cell Res 260:308–313

Chapter 13

Fast Micromethod: Determination of DNA Integrity in Cell Suspensions and in Solid Tissues

Nevenka Bihari

Abstract

The Fast Micromethod is a rapid and convenient microplate procedure for the determination of DNA integrity in cell suspensions and in solid tissues. The procedure is based on the ability of fluorochromes to preferentially interact with double-stranded DNA in alkaline conditions. Rapid sample lysis is followed by denaturation at high pH during 15 min. Only 30 ng of DNA from cell suspensions or tissue homogenates per single well are required for the analyses. The whole analysis is performed within 3 h or less (for one 96-well microplate).

The Fast Micromethod is broadly used in biology and medicine. Its applications range from environmental pollution tests in marine invertebrates to the analysis of biopsy samples in cancer patients to detect DNA alterations caused by irradiation or chemotherapy.

The procedure presented here describes the Fast Micromethod applied for the determination of DNA integrity in cell suspensions (HeLa cells) and solid tissues (mussel gills).

Key words Genotoxicity assessment, Rapid determination of DNA integrity, Environmental pollutants, Mussel gills, Cell suspensions

1 Introduction

The Fast Micromethod measures reduced DNA integrity caused by genotoxins in investigated samples in comparison to a high-integrity DNA in reference samples. Alkaline-labile sites and DNA single- or double-strand breaks are the primary DNA lesions that significantly reduce DNA integrity. The method measures the rate of cellular DNA unwinding upon exposure to alkaline conditions. The assumption is that hydrogen bonds in double-stranded DNA are destabilized and the two strands separated in high alkaline and high ionic solutions. Low-integrity DNA strands separate faster than high-integrity DNA strands. In order to distinguish between low and high-integrity DNA, it is convenient to use a fluorescent dye (PicoGreen) that preferentially binds to double-stranded DNA but not to single-stranded DNA or proteins. The measurement of the

Vladimir V. Didenko (ed.), *Fast Detection of DNA Damage: Methods and Protocols*, Methods in Molecular Biology, vol. 1644, DOI 10.1007/978-1-4939-7187-9_13, © Springer Science+Business Media LLC 2017

dye-DNA complex allows the determination of the kinetic pattern of double-stranded DNA unwinding. The kinetic pattern of the double-stranded DNA unwinding depends on the length of the molecule, but also on the number of single-strand breaks and alkali-labile sites present in the DNA. DNA with single-strand breaks and alkali-labile sites unwinds faster than DNA that is not damaged. The use of the PicoGreen dye (which preferentially stains double-stranded DNA) allows the quantification of double-stranded DNA which remains unwound. Thus, the Fast Micromethod can determine the frequency of DNA single-strand breaks.

The Fast Micromethod for the determination of DNA integrity in cell suspensions and solid tissues was first described in a patent application in 1997 [1]. The altered DNA integrity induced by γ-radiation was tested in mouse lymphoblasts, human lymphocytes, and mouse tissues (liver and muscle) [2]. The sensitivity of the method is comparable with the Comet assay for the determination of DNA integrity in HeLa cells induced by γ-radiation, UV-C light, and the chemical agent 4-nitroquinolone-*N*-oxide [3]. The technique overview was reported in 2006 [4] and more recently in 2008 [5].

The Fast Micromethod is a simple technique used routinely for the assessment of DNA integrity for medical purposes, for genotoxicity assessments in biomonitoring programs and the assessment of ecosystem conditions. A decrease in DNA integrity was detected in irradiated human peripheral blood mononuclear cells (PBMC) [6] and in isolated PBMCs from cancer patients after radiotherapy [7]. The Fast Micromethod was the method of choice in extensive studies of marine sponge *Suberites domuncula* exposed to polycyclic aromatic hydrocarbons [8] or cadmium chloride [9], in UV-B irradiated marine sponge *Geodia cydonium* [10] as well as in the sponge *Baikalo spongia intermedia* exposed to heavy metals [11]. The DNA damage (low DNA integrity) was also examined as a biomarker in the mussels *Mytilus galloprovincialis* sampled at differently contaminated areas of the Istrian coast, Northern Adriatic, Croatia [12]. Alterations of DNA integrity in mussel gills were detected in 40% of the 240 mussels collected along the Adriatic coast [13]. A long-term field study of DNA integrity in mussel gills enabled us to define "hot spots" as sites with the presence of different genotoxin loads in the marine environment [14, 15]. Additionally, using this technique to determine DNA integrity in the liver of the fish *Limanda limanda* confirmed it as a convenient method for environmental monitoring studies [16].

This chapter describes the Fast Micromethod procedure for the determination of DNA integrity in cell suspensions (HeLa cells) and solid tissues (mussel gills). The procedure includes (1) cell suspensions preparation, (2) tissue sample preparations, (3) determination of DNA content in solid tissues, (4) simple and rapid sample lysis, (5) DNA integrity measurements, (6) fluorescence analysis and calculation of strand scission factors, and (7) results presentation.

2 Materials

1. Sample buffer: 10% DMSO solution in TE buffer (1 mM EDTA, 10 mM Tris–HCl, pH 7.4).
2. YOYO-1-iodide working solution: separate the original stock of 1 M YOYO-1-iodide in 5 μL aliquots and store at −20 °C (*see* **Note 1**). Immediately before analysis, add 500 μL of sample buffer to the 5 μL YOYO-1-iodide aliquot in order to get a 4×10^{-7} M working solution.
3. Lysing solution: 4.5 M urea, 0.2 M EDTA, 0.1% SDS, pH 10.0 supplemented with 20 μL of the original stock dye PicoGreen in 1 mL of lysing solution.
4. Working alkaline solution: 0.1 M NaOH in 20 mM EDTA of a specific pH (*see* **Note 2**).
5. Reaction mixture: 25 μL dissolved tissue homogenate, 25 μL lysing solution, 250 μL alkaline working solution.

3 Method

The critical point for the determination of DNA integrity in solid tissues with Fast Micromethod is to preserve the DNA from any additional damage caused by handling procedures. Therefore, after its dissection the solid tissue has to be immediately frozen using liquid nitrogen. This step is especially important for the transportation of collected in situ samples (e.g., gills of mussels from different sampling sites). Upon arrival to the laboratory, the frozen samples can be stored in liquid nitrogen for 1 year.

3.1 Cell Suspensions Preparation

1. Use exponentially growing cells. Cultures of human HeLa cells are maintained in RPMI 1640 medium with 10 mM *N*-2-hydroxyethylpiperazine-*N′*-2-ethanesulfonic acid (HEPES) and 10% FCS. Count the cells.
2. Place exactly the same amount of cells in each single well of the microplate: 3000 cells/25 μL of Ca/Mg-free PBS or TE buffer.

3.2 Tissue Sample Preparations

1. Precool a mortar and a pestle in liquid nitrogen. Homogenize 100 mg of tissue samples with 2 mL of sample buffer in a mortar with liquid nitrogen.
2. Collect the pellets in test tubes and store at −80 °C (*see* **Note 3**).

3.3 Determination of Total DNA Content in Tissue Homogenates

1. Dilute one aliquot of the tissue homogenate with the sample buffer. For tissue homogenates the desired DNA concentration is usually 100 ng/mL (*see* **Note 4**) and for serial standard DNA solutions the desired concentrations are 50–500 ng/mL.

2. For the determination of total DNA content in tissue homogenates use the YOYO-1-iodide dye (*see* **Note 5**). In a fluorometric microplate mix 50 μL of the diluted sample homogenate or 50 μL of the DNA standard solutions with 50 μL of the YOYO working solution (*see* **Note 6**).
3. Measure the fluorescence at E_{exc} 485 nm and E_{ems} 520 nm in the fluorometric microplate reader for 5 min. Calculate the total DNA content in the samples in comparison with the standard DNA curve.

3.4 Sample Lysis

1. Mix 25 μL of cell culture (3000–5000 cells) or 25 μL tissue homogenate suspension (25 μg of tissue, cca 30–50 ng DNA) with 25 μL of lysing solution in the microplates. Prepare blank sample by mixing 25 μL of cell culture medium, or 25 μL tissue homogenate buffer with 25 μL of lysing solution in the microplates.
2. Keep the microplates in the dark and on ice for 40 min (*see* **Notes** 7 and **8**).

3.5 DNA Integrity Measurement

1. Initiate the alkaline denaturation of double-stranded DNA from the lysed tissue or cells in the microplate wells with the addition of 250 μL working alkaline solution.
2. Immediately place the microplate in the fluorometric microplate reader and follow DNA denaturation directly, for 15 min every 30 s, by measuring the fluorescence of the double-stranded DNA-PicoGreen complex at E_{exc} 485 nm and E_{ems} 520 nm (*see* **Note 9**).

3.6 Fluorescence Analysis and Calculation of Strand Scission Factors

1. Correct the recorded fluorescence values for blank readings. Correct for blank readings by subtracting the blank values from all the values.
2. Calculate the average fluorescence unit values of the double-stranded DNA-PicoGreen complex fluorescence quadruplicate samples and draw the denaturation curves.
3. Calculate the strand scission factor (SSF) 7 min (*see* **Note 10**) after denaturation according to the equation:

$$SSF = \log 10\left(\text{Fluorescence Units}_{\text{sample}} / \text{Fluorescence Units}_{\text{reference}}\right)\left(\textit{see}\ \textbf{Note 11}\right)$$

Thus SSF = 0 reveals high DNA integrity (no difference between reference and sample DNA integrity), while SSF < 0 indicates loss of DNA integrity in samples.

3.7 Results Presentation

Express the results as strand scission factors (SSFs) with corresponding standard deviations (*see* **Note 12**) and present as histograms or scatter plot diagrams. Histograms are preferable for the presentation of the analyzed data on the biological effects of genotoxins. Due to the assumption that DNA integrity is lower in the investigated samples as compared to the reference sample (SSF < 0), it is practical to multiply the SSFs values by (−1) and then present them as histograms. In such a way, the bars representing the sample histograms will be higher than those representing the control-reference sample. The higher are the histogram bars the lower is the DNA integrity of the analyzed samples revealing a greater biological effect of the investigated genotoxin.

4 Notes

1. Expected shell life of YOYO dye is 6–12 months. All YOYO solutions have to be stored in plastic vials at −20 °C and protected from light.
2. Prepare the working alkaline solution daily in order to assure the specific final pH of the reaction mixture. Mix 2.5 mL of working alkaline solution with 0.25 mL TE buffer (1 mM EDTA, 10 mM Tris–HCl, pH 7.4) and 0.25 mL lysing solution and check the pH. If necessary, add a few more drops of the alkaline working solution or 20 mM EDTA solution, make a new reaction mixture and check the pH. Repeat the procedure until the proper pH is reached. pH adjustment is an essential step in order to achieve proper DNA denaturation in alkaline conditions due to different DNA lengths and complexity in different cell types or organism. The pH has to be 12.4 for human lymphocytes, 11.6 for RTG-2 cell lines, and 11.5 for mussel gills DNA. For any other cell lines or tissues use the pH at which the difference between the slopes of the denaturation curves for reference and investigated samples is the highest.
3. Tissue samples can be stored at −80 °C for 1 year.
4. For mussel gills we usually dilute dissolved samples tenfold and obtain 80–130 ng/mL DNA.
5. The total DNA content in samples may be assessed by UV measurement of the 260 nm/280 nm ratio, but the fluorochromatic dye YOYO-1-iodide provides a more sensitive and precise determination of the total DNA content in the presence of other macromolecules in crude tissue homogenates or cell suspensions. For Fast Micromethod the same quantity of DNA should be placed in each well of the microplate. This can be achieved by a precise measurement of DNA content in solid tissues or cells and its standardization to 100 ng/mL of tissue DNA or 10^5 cells. Therefore, the YOYO®-1-iodide determination of the DNA content is not only recommended but mandatory prior to the Fast Micromethod analyses.

6. The final YOYO-1-iodide concentration is 2×10^{-7} M as recommended [17].
7. The interactions between the fluorochrome dye PicoGreen and DNA occur during lysing. The temperature, buffer component, and time are important factors during the lysing period. Lysing with 9 M urea solution on ice is recommended in order to suppress DNA repair mechanisms [2]. However, this may lead to the precipitation of the lysing buffer components (e.g., urea). Therefore, the urea concentration has to be reduced to 4.5 M and the lysing step has to be performed at room temperature for only 30 min [18].
8. Each sample has to be analyzed in quadruplicates.
9. The PicoGreen fluorochrome exhibits a 1000-fold increase in the fluorescence yield upon binding to double-stranded DNA compared to the free fluorochrome in the solution. A simple approach to quantifying the relative quantity of DNA in solutions containing both double-stranded DNA and single-stranded DNA is possible due to different lifetime characteristics of each of the DNA-fluorochrome complexes (i.e., double-stranded DNA–PicoGreen and single-stranded DNA–PicoGreen) [19]. The fluorescence values of calf thymus single-stranded DNA–PicoGreen and double-stranded DNA–PicoGreen complex show large differences between the two DNA-fluorochrome complexes over a wide range of DNA concentrations (1–33 ng DNA) and alkaline conditions (pH 7.5–12.6).
10. The DNA denaturation time for the calculation of SSF is defined by the differences of the slopes of the denaturation curves for reference and investigated samples. It is the time interval where the difference of the slopes is constant. In the case of mussel gills DNA this time is 7 min, for mouse liver 10 min and for human liver 15 min. The slopes of the denaturation curves describe the rate of DNA denaturation and the decay of the double-stranded DNA-PicoGreen complexes in alkaline conditions. The rate of DNA denaturation is calculated as the difference in fluorescence values for the desired time interval (time 1 and time 2) divided by the denaturation time:

$$\text{Slope} = \Delta\,\text{Fluorescence Units} / \text{time}\,(\text{min})$$

11. Equation derived from the definition of the strand scission factor [20]

$$SSF = \log 10\left(\%\,\text{dsDNA}_{\text{sample}} / \%\,\text{dsDNA}_{\text{reference}}\right)$$

Thus SSF = 0 indicates an absence of additional DNA strand breaks and alkaline-labile sites, while SSF < 0 indicates increasing frequencies of strand breaks and alkaline-labile sites in the samples.

12. Standard deviations are relatively high in experiments determining DNA integrity in mussel gills. In the case of mussel samples the standard deviation is derived from quadruplicate measurements of no less than five mussels per concentration of a tested genotoxin or, at least, from five in situ collected mussels.

When there is a significant statistical difference between tissue samples and the reference sample, it is possible to conclude that there is indeed a biological effect of the investigated genotoxin on DNA integrity in mussel gills cells. If the statistical difference is not significant enough, it is not possible to determine the biological effect of the investigated genotoxin on DNA integrity. This can potentially be related to repair processes and/or presence of cross-links.

References

1. Müller WEG, Batel R, Zahn R et al (1997) Mikro-methode zur Schnellbestimmung von DNA-Schäden und deren Reparatur unter Verwendungs des Fluoreszenz farbstoffs PicoGreen und ihre Anwendung. Deutche Patenteamt Muenchen. No. 19724781.4-41
2. Batel R, Jaksić Ž, Bihari N et al (1999) A microplate assay for DNA damage determination (Fast Micromethod) in cell suspensions and solid tissues. Anal Biochem 270:195–200
3. Bihari N, Batel R, Jakšić Ž et al (2002) Comparison between the comet assay and Fast Micromethod® for measuring DNA damage in HeLa cells. Croatica Chem Acta 75:793–804
4. Schröder HC, Batel R, Schwertner H et al (2006) Fast Micromethod DNA single-strand-break assay. In: Henderson DS (ed) DNA repair protocols: mammalian systems. Humana Press Inc, Totowa
5. Jakšić Ž (2008) Assessing the Fast Micromethod as a tool in DNA integrity measurement. In: Souta M, Shouta N (eds) Progress in DNA damage research. Nova Publishers, New York
6. Schröder HC, Chauvin C, Lauenroth S et al (1998) Comparison of frequency of apoptosis in X-irradiated human peripheral blood mononuclear cells from donors of various ages. In: Schröder HC, Müller WEG (eds) Development of methods for determination of radiosensitivity. Akademie Gemeinnutziger Wissenschaften, Mainz
7. Elmendorff-Dreikorn K, Chauvin C, Muller WEG et al (1998) Comparison of frequency of apoptosis in X-irradiated human peripheral blood mononuclear cells from donors of various ages. In: Schröder HC, Müller WEG (eds) Development of methods for determination of radiosensitivity. Akademie Gemeinnutziger Wissenschaften, Mainz
8. Schröder HC, Batel R, Lauenroth S et al (1999) Induction of DNA damage and expression of heat shock protein HSP70 by polychlorinated biphenyls in the marine sponge *Suberites domuncula* Olivi. J Exp Mar Biol Ecol 233:285–300
9. Schröder HC, Hassanein HMA, Lauenroth S et al (1999) Induction of DNA strand breaks and expression of HSP70 and GRP78 homolog by cadmium in the marine sponge *Suberites domuncula*. Arch Environ Contam Toxicol 36:47–55
10. Batel R, Fafandjel M, Blumbach B et al (1998) Expression of the human XPB/ERCC-3 excision repair gene-homolog in the sponge *Geodia cydonium* after exposure to ultraviolet radiation. Mutat Res 409:123–133
11. Efremova SM, Margulis BA, Guzhova IV et al (2002) Heat shock protein Hsp70 expression and DNA damage in Baikalian sponges exposed to model pollutants and wastewater from Baikalsk Pulp and Paper Plant. Aquat Toxicol 57:267–280

12. Bihari N, Hamer B, Jakšić Ž et al (2002) Application of alkaline elution, Fast Micromethod and flow cytometry in detection of marine contamination. Cell Mol Biol 48(4):373–377
13. Bihari N, Mičić M, Fafanđel M et al (2004) Testing quality of sea water from the Adriatic coast of Croatia with toxicity, genotoxicity and DNA integrity tests. Croatica Chem Acta 45(1):75–81
14. Jakšić Ž, Batel R, Bihari N et al (2005) Adriatic coast as a microcosm for global genotoxic marine contamination—a long-term field study. Mar Pollut Bull 50:1314–1327
15. Bihari N, Fafanđel M, Jakšić Ž et al (2005) Spatial distribution of DNA integrity in mussels, *Mytilusgalloprovincialis*, from the Adriatic sea, Croatia. Bull Environ Contam Toxicol 75:845–850
16. Schröder HC, Batel R, Hassanein HMA et al (2000) Correlation between the level of the potential biomarker, heatshock protein, and the occurrence of DNA damage in the dab, *Limanda limanda*: a field study in the North Sea and the English Channel. Mar Environ Res 49:201–215
17. Rye HS, Dabora JM, Quesada MA et al (1993) Fluorometric assay using dimeric dyes for double and single-stranded DNA and RNA with picogram sensitivity. Anal Biochem 208:144–150
18. Jakšić Ž, Batel R (2003) DNA integrity determination in marine invertebrates by Fast Micromethod®. Aquat Toxicol 65:361–376
19. Cosa G, Foscaneanu KS, McLean JRN et al (2001) Photophysical properties of fluorescent DNA—dyes bound to single- and double-stranded DNA in aqueous buffered solution. Photochem Photobiol 73:585–599
20. Meyen RE, Jenkins WT (1983) Variation in normal and tumor tissue sensitivity of mice to ionising radiation induced DNA strand breaks in vivo. Cancer Res 43:5668–5673

Chapter 14

Quantification of DNA Damage and Repair in Mitochondrial, Nuclear, and Bacterial Genomes by Real-Time PCR

Nadja Patenge

Abstract

DNA damage caused by genotoxic insults is often used as an indicator of specific diseases, environmental challenges, and metabolic processes. To date, various different methods have been described to detect damaged DNA. Many techniques need high amounts of DNA for the analysis and/or require the exact determination of DNA template concentration. Here, we describe a rapid and quantitative method for the evaluation of the relative levels of damage in mitochondrial, nuclear, and bacterial DNA in comparison to untreated controls. The approach is based on the real-time PCR amplification of DNA fragments of two different lengths in the respective samples. DNA damage detection using this protocol is gene-specific. The technique can also be expanded to monitor DNA repair and to detect genomic hot-spots for DNA lesions.

Key words DNA damage, Quantitative PCR, Real-time PCR, DNA repair, Detection of DNA damage, Genomic integrity, Mitochondrial DNA, Mitochondrial maintenance

1 Introduction

Metabolic processes, disease, UV-radiation, drugs, and industrial waste cause the formation of reactive oxygen species (ROS) in living organisms and lead to DNA damage. Quantitative determination of DNA damage under specific conditions is important for the investigation of infection and chronic disease, for studying aging and mitochondrial DNA maintenance. DNA damage is an indicator of drug toxicity and an important marker used in ecotoxicology. Furthermore, DNA integrity serves as a parameter for the evaluation of the quality of stem cells or cryopreserved sperm.

A broad spectrum of approaches is available to detect damage in prokaryotic as well as eukaryotic DNA. Typically used methods include Southern blot analysis, which is able to detect DNA strand breaks semi-quantitatively but requires high amounts of sample material [1]. The Comet assay is based on single cell gel electrophoresis. It is a widely used approach detecting single- and

Vladimir V. Didenko (ed.), *Fast Detection of DNA Damage: Methods and Protocols*, Methods in Molecular Biology, vol. 1644, DOI 10.1007/978-1-4939-7187-9_14, © Springer Science+Business Media LLC 2017

double-strand breaks as well as alkali-labile DNA sites under alkaline conditions [2]. In contrast, high-performance liquid chromatography (HPLC) in combination with different detection methods allows quantitative detection of a subset of specifically damaged DNA products, e.g., 8-oxoguanine (8-oxoG) or abasic sites [3].

A long-run quantitative PCR method was developed to minimize the required input material. It permits the amplification of 25 kb long DNA fragments using just nanogram amounts of total DNA [4]. In this method, band density of the PCR product is quantified either by Southern blotting or by binding of a fluorescent dye. Thus, the multi-step procedure is still time-consuming, and requires a high degree of optimization and accuracy for reliable experimental results. Here, we describe a rapid, real-time, quantitative PCR (qPCR) method for the accurate, relative quantification of DNA lesions [5]. The approach requires 3 h to complete.

DNA lesions caused by genotoxic insults include double- and single-strand breaks, abasic sites, and DNA adducts. These DNA alterations are known to inhibit primer annealing or to block polymerase-based DNA synthesis [6]. In the described qPCR approach, the disruption of DNA polymerization is utilized for the quantitative determination of DNA damage in a stressed sample in comparison to an untreated control. In this method each template is amplified in two independent reactions with distinct primer pairs.

The first primer pair leads to the production of a short amplicon. This reaction serves as an internal concentration control, because the probability for lesion formation within the small region is low. For the small amplicon, cycle threshold (Ct) values of the treated sample and of the undamaged control are very similar (Fig. 1a). In the second reaction, a long amplicon is produced, which is more efficiently synthesized in the untreated sample. Upon induction of DNA damage, a number of template molecules will undergo DNA modifications, which disrupt DNA polymerization or delay polymerase progression. As a consequence, the Ct-value of the treated sample will be higher in comparison to the undamaged control sample (Fig. 1b). Differential Ct-value determination allows the calculation of lesions per template. Comparison of a small amplicon (internal concentration control) to a long amplicon (sample) allows accurate DNA quantification without extensive adjustment of DNA input concentration between samples. Furthermore, if working with mtDNA, normalization for mitochondrial copy number is achieved by this approach [7].

Initially, the method was developed for the detection of ROS-induced mtDNA damage in cultured cells [2, 7]. Currently, we use this technique for damage detection in bacterial DNA, and in mtDNA extracted from mouse tissue samples for the monitoring of the infection processes. Several laboratories have used this protocol to investigate the role of eukaryotic DNA damage in diabetes, cancer, and neurodegeneration [8–12]. Furthermore, the method was applied to test the quality of cryopreserved human

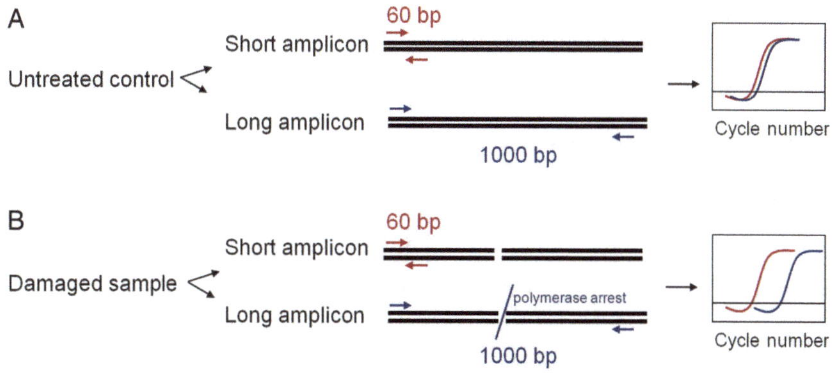

Fig. 1 Schematic of the assay principle. (**a**) PCR reactions with the untreated template result in a short and a long product, depending on the primer pair employed. Due to the comparable efficiency of these reactions, the Ct values of the fluorescence in the two reactions are similar. (**b**) When the damaged sample is used as a template in PCR, the amplification of the short amplicon differs from the amplification of the long amplicon. The probability for lesion generation is low within the short fragment compared to the long fragment. Thus, PCR reactions with the primer pair specific for the short fragment tend to proceed without disturbances. In contrast, in reactions with primers specific for the long fragment, the DNA lesions block polymerase activity in a dose-dependent manner in a proportion of the template molecules. This increases the Ct values as compared to the short fragment

and fish sperm [13–15]. Improved polymerases and dyes allow the use of longer amplicons and thereby increase sensitivity of this technique [16, 17].

2 Materials

2.1 DNA Isolation

1. DNeasy Blood & Tissue Kit (Qiagen) (*see* **Note 1**).
2. UV/Vis Spectrophotometer.

2.2 qPCR

1. PCR primers. All primers for real-time PCR listed here were designed using Primer 3 software (*see* **Note 2**). PCR primers should be HPLC-purified. The sequences of all the primers used in this protocol are listed in Table 1 (*see* **Note 3**).
2. Maxima SYBR Green/ROX qPCR Master Mix (*see* **Note 4**).
3. PCR 96-well microtiter plates for LC 480 II.
4. Roche LightCycler 480 II Real Time PCR System (*see* **Note 5**).
5. Roche LightCycler® Software 4.0.

3 Methods

3.1 DNA Isolation

1. Collect samples from an experiment under the conditions of interest, e.g. exposure to oxidative stressors, or from environmental specimen. Any biological sample from which total

Table 1
Oligonucleotides for DNA damage determination

Amplicon name	Forward strand (5′ → 3′)	Amplicon length (bps)	PCR efficiency (%)
S. pyogenes, *gyrA* long		775	94
Forward strand (5′-3′)	GGGATGGCAACTAACATTCCG		
Reverse strand (5′-3′)	TCCAAACCAGTCAAACGACGC		
S. pyogenes, *gyrA* short		89	95
Forward strand (5′-3′)	CAGACAGAATTGATGTCACG		
Reverse strand (5′-3′)	TCCAAACCAGTCAAACGACGC		
mmtD-loop[a] long		618	73
Forward strand (5′-3′)	CCATATGACTATCCCCTTCC		
Reverse strand (5′-3′)	GATTAGAGTTTTGGTTCACGG		
mmtD-loop short		69	92
Forward strand (5′-3′)	AAGGACAGCACACAGTCTAG		
Reverse strand (5′-3′)	CTTAGGTGATTGGGTTTTGCG		
hmtD-loop[b] long		972	78
Forward strand (5′-3′)	CTGTTCTTTCATGGGGAAGC		
Reverse strand (5′-3′)	AAAGTGCATACCGCCAAAAG		
hmtD-loop[b] short		55	98
Forward strand (5′-3′)	CCCTAACACCAGCCTAACCA		
Reverse strand (5′-3′)	AAAGTGCATACCGCCAAAAG		

[a]mmtD-loop: mouse mitochondrial D-loop
[b]hmtD-loop: human mitochondrial D-loop

DNA can be isolated is suitable for the analysis. These include bacterial pellets, homogenized organ tissues, and cultured cells. A group of nontreated samples should serve as controls.

2. Purify total DNA from your sample (*see* **Note 1**).
3. Determine DNA quantity and purity by spectrometric analysis. The DNA should exhibit a high purity (A260/A280 > 1.8) and can be stored at 4 °C for at least a month.

3.2 qPCR Reaction

1. Set up a reaction mix per well containing: 10 μL Maxima SYBR Green/ROX qPCR Master Mix (2X), 1.2 μL of forward primer (10 μM), 1.2 μL reverse primer (10 μM), DNA sample (2–100 ng) (*see* **Note 6**), and nuclease-free water (included in the SYBR Green kit) to 20 μL. Perform each reaction in triplicates. Include a non-template control. Prepare a master mix whenever appropriate to minimize pipetting errors.

2. Due to the different cycling conditions for long and short fragments, carry out the respective reactions in separate 96-well plates. For short amplicons use the following cycling conditions: 10 min at 95 °C, 10 s at 95 °C, 10 s at 60 °C, 10 s at 72 °C, 40 cycles. Melting curve: 5 s at 95 °C, 15 s at 70 °C.

   For long amplicons use the following program: 10 min at 95 °C, 10 s at 95 °C, 10 s at 60 °C, 30 s at 72 °C, 40 cycles. Melting curve: 5 s at 95 °C, 15 s at 70 °C.

3. Calculate Ct values by using the LightCycler® Software 4.0.

3.3 Determination of PCR Reaction Efficiencies

The purpose of this step is to evaluate if the primers designed for the respective target region are suitable for relative quantification of the template DNA. Furthermore, at this stage you also determine the DNA concentration range, which can be used for quantification. Within the appropriate concentration range, the amplification will be linear with respect to the input DNA concentration.

1. Set up serial dilutions of total DNA from an untreated sample up to 1 : 10,000, corresponding to a DNA concentration range of approximately 50–0.005 ng/μL (*see* **Note 7**).
2. Perform a qPCR employing the conditions described in Subheading 3.2. qPCR reaction. Use each of the different dilutions as a template for triplicate reactions with the primer pair to be tested.
3. Plot the mean of the Ct values of each triplicate against the quantity of template (e.g., dilutions 10^{-5}–10^{-1}) to create a standard curve.
4. Calculate PCR efficiencies using the linear regression slope of the standard curve using the following formula: efficiency = $-1 + 10^{(-1/\text{slope})}$ (*see* **Note 8**). Slopes between −3.3 and −4.0 correspond to reaction efficiencies between 100% and 78%.

3.4 DNA Lesion Rate Analysis

The DNA lesion rate is estimated on the basis of the assumption of an even distribution of the damaged sites. The underlying analyses follow the $2^{-\Delta\Delta Ct}$ method [18].

1. Perform qPCR reactions using the conditions described in Subheading 3.2. qPCR reaction. Use DNA purified from non-treated and treated samples as templates.
2. For each sample, set up independent triplicate reactions employing primers specific for the short amplicon and primers specific for the long amplicon, respectively.
3. Use the mean of the Ct values of each triplicate for the calculation of the lesion rate (*see* **Note 9**).
4. Compare each of the treated samples individually to the non-treated control. For this first determine differences between

the mean Ct values of treated samples and nontreated samples for the long amplicon. Then repeat the procedure with the mean Ct values of treated samples and nontreated samples for the short amplicon. Calculate the lesion rate per 10 kb according to the following formula (*see* **Note 10**):

$$\text{Lesion rate} = 1 - 2^{-(\Delta Ct\ \text{long} - \Delta Ct\ \text{short})} \times \frac{10{,}000\,(bps)}{\text{Size of long fragment}\,(bps)}$$

4 Notes

1. If you have established a protocol to prepare high-quality DNA from your sample, there is no need to change your system. Low amounts of pure DNA can be used as input material regardless of the purification protocol. However, DNA quality for qPCR should be tested with a set of known primers before starting the experiment. Make sure that your DNA purification protocol does not damage your sample DNA, e.g., by shearing.
2. The detection method is gene-specific. Design your own primers specific for your organism of interest and appropriate for your objective. Besides testing the DNA damage in the sample cells, the technique can be used to find hot spots of DNA lesion formation in the genome following exposure to a specific stressor. It is crucial to test the PCR efficiency for your specific primer-template combination. It may be challenging to establish a high PCR efficiency for the long amplicons. Be sure to use a set of primers that allow an efficiency of at least 70%.
3. More examples of useful primers can be found in the publications cited in the Introduction to this chapter. Several working groups used this protocol with primers specific for different organisms.
4. The qPCR reaction works with intercalating dyes as well as with labeled oligonucleotide probes. We used intercalating dyes to work with several primer sets in studies of different organisms and genes. If you are planning to use a limited set of primers, it might be worthwhile establishing your reactions using fluorophore-labeled oligonucleotides [19].
5. There is no need to change the cycler or the reagents if a different type of PCR cycler and an established qPCR protocol is already in use in the laboratory. Adapt the cycle conditions for the long amplicon and make sure to verify the conditions and the primer suitability by determining the reaction efficiency as described in Subheading 3.3.

6. The optimal DNA amount depends on the complexity of the sample and the efficiency of the primers which is determined as described in Subheading 3.3.
7. The concentration range to be tested should include the anticipated DNA concentration expected to be attained from the experimental samples. This permits determining if the qPCR-based analysis is sensitive enough for the amount of DNA available from the actual samples.
8. Conveniently, there are several free online PCR efficiency calculators available.
9. The PCR reaction should be redone if one of the Ct values from the triplicate reactions differs noticeably from the other two, or if all three Ct values are inconsistent. Do not attempt to use the mean of variable Ct values for the lesion analyses. This will interfere with a thorough data interpretation.
10. Improved polymerases and dyes can increase sensitivity of the technique because they permit using longer amplicons. For the calculation of DNA damage on long amplicons >1000 bp, another algorithm can be employed [20]. To achieve a higher accuracy, the reaction efficiency can be considered in the analyses [14, 17].

References

1. Niwa O (2003) Induced genomic instability in irradiated germ cells and in the offspring; reconciling discrepancies among the human and animal studies. Oncogene 22:7078–7086
2. Singh NP (2016) The comet assay: reflections on its development, evolution and applications. Mutat Res Rev 767:23–30
3. Taghizadeh K, McFaline JL, Pang B et al (2008) Quantification of DNA damage products resulting from deamination, oxidation and reaction with products of lipid peroxidation by liquid chromatography isotope dilution tandem mass spectrometry. Nat Protoc 3:1287–1298
4. Santos JH, Meyer JN, Mandavilli BS et al (2006) Quantitative PCR-based measurement of nuclear and mitochondrial DNA damage and repair in mammalian cells. Methods Mol Biol 314:183–199
5. Rothfuss O, Gasser T, Patenge N (2010) Analysis of differential DNA damage in the mitochondrial genome employing a semi-long run real-time PCR approach. Nucleic Acids Res 38:e24
6. Sikorsky JA, Primerano DA, Fenger TW et al (2007) DNA damage reduces Taq DNA polymerase fidelity and PCR amplification efficiency. Biochem Biophys Res Commun 355:431–437
7. Rothfuss O, Fischer H, Hasegawa T et al (2009) Parkin protects mitochondrial genome integrity and supports mitochondrial DNA repair. Hum Mol Genet 18:3832–3850
8. Artuso L, Zoccolella S, Favia P et al (2013) Mitochondrial genome aberrations in skeletal muscle of patients with motor neuron disease. Amyotroph Lateral Scler Frontotemporal Degener 14:261–266
9. Fendt L, Niederstatter H, Huber G et al (2011) Accumulation of mutations over the entire mitochondrial genome of breast cancer cells obtained by tissue microdissection. Breast Cancer Res Treat 128:327–336
10. Kowluru RA (2013) Mitochondria damage in the pathogenesis of diabetic retinopathy and in the metabolic memory associated with its continued progression. Curr Med Chem 20:3226–3233
11. Saydoff JA, Liu LS, Garcia RA et al (2003) Oral uridine pro-drug PN401 decreases neurodegeneration, behavioral impairment, weight loss and mortality in the 3-nitropropionic acid mitochondrial toxin model of Huntington's disease. Brain Res 994:44–54

12. Tewari S, Santos JM, Kowluru RA (2012) Damaged mitochondrial DNA replication system and the development of diabetic retinopathy. Antioxid Redox Signal 17:492–504
13. Carton-Garcia F, Riesco MF, Cabrita E et al (2013) Quantification of lesions in nuclear and mitochondrial genes of *Sparus aurata* cryopreserved sperm. Aquaculture 402-403:106–112
14. Gonzalez-Rojo S, Fernandez-Diez C, Guerra SM et al (2014) Differential gene susceptibility to sperm DNA damage: analysis of developmental key genes in trout. PLoS One 9:e114161
15. Valcarce DG, Carton-Garcia F, Riesco MF et al (2013) Analysis of DNA damage after human sperm cryopreservation in genes crucial for fertilization and early embryo development. Andrology 1:723–730
16. Edwards JG (2009) Quantification of mitochondrial DNA (mtDNA) damage and error rates by real-time QPCR. Mitochondrion 9:31–35
17. Lehle S, Hildebrand DG, Merz B et al (2013) LORD-Q: a long-run real-time PCR-based DNA-damage quantification method for nuclear and mitochondrial genome analysis. Nucleic Acids Res 42:e41
18. Livak KJ, Schmittgen TD (2001) Analysis of relative gene expression data using real-time quantitative PCR and the 2(−Delta Delta C(T)) method. Methods 25:402–408
19. Navarro E, Serrano-Heras G, Castano MJ et al (2015) Real-time PCR detection chemistry. Clin Chim Acta 439:231–250
20. Furda A, Santos JH, Meyer JN et al (2014) Quantitative PCR-based measurement of nuclear and mitochondrial DNA damage and repair in mammalian cells. Methods Mol Biol 1105:419–437

Chapter 15

Zebra Tail Amplification: Accelerated Detection of Apoptotic Blunt-Ended DNA Breaks by In Situ Ligation

Vladimir V. Didenko

Abstract

In situ ligation (ISL) is a simple and specific technique for apoptosis labeling in tissue sections. In its most economical version ISL uses ordinary PCR-labeled DNA fragments as probes. In tissue sections these makeshift probes are ligated to apoptotic DNA breaks by T4 DNA ligase. The approach can selectively label 5′PO_4 DNA breaks with blunt ends, and is the histological equivalent of electrophoretic apoptotic ladder detection. The main drawback of this technique is its low speed, as it requires 18 h-incubation for efficient labeling. Here, we describe an easy modification of ISL which reduces the incubation time to 1 h and converts ISL into a rapid detection method taking ~3 h overall. Signal enhancement is achieved by a new type of isothermal amplification reaction which generates "zebra tails"— long and labeled extensions of the probes attached to DNA breaks.

Key words In situ ligation, Zebra tail amplification, Rapid apoptosis detection, Isothermal amplification, Blunt-ended DNA breaks labeling, Signal enhancement

1 Introduction

The majority of DNA fragments produced in apoptotic DNA cleavage are blunt-ended and have 3′OH/5′PO_4 functional groups at the ends [1–3]. The techniques that simultaneously detect several of these DNA cleavage features can label apoptotic cells with increased specificity. In situ ligation (ISL) is one of those techniques [4]. This approach detects two markers of apoptotic DNA fragmentation: blunt DNA ends and terminal 5′ phosphates in DNA breaks in fixed apoptotic cells. The reaction is carried out by adding blunt-ended 5′OH DNA probes and T4DNA ligase directly to tissue sections [5] (Fig. 1). The ISL labeling occurs only when both markers of apoptotic DNA cleavage are simultaneously present. As a result, the assay is more specific for apoptosis than TUNEL which labels a single marker of broken DNA—free 3′OH groups [4, 6–8]. Free 3′OH groups in DNA represent a much more generic type of DNA damage and can be produced as artifacts of

Vladimir V. Didenko (ed.), *Fast Detection of DNA Damage: Methods and Protocols*, Methods in Molecular Biology, vol. 1644,
DOI 10.1007/978-1-4939-7187-9_15,

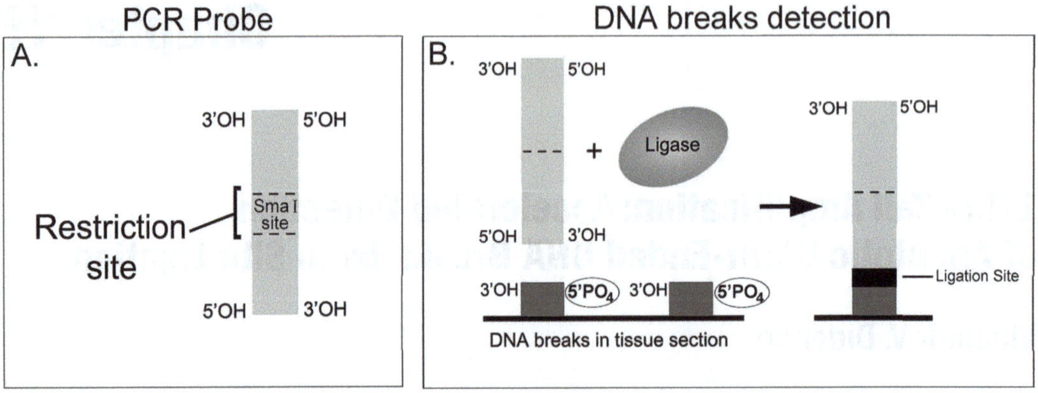

Fig. 1 ISL probe and regular ISL labeling. (**a**) A regular ISL probe is a blunt-ended DNA fragment with 3′OH/5′OH at the ends. It can be tagged with either digoxigenin or fluorophores. The probe's actual sequence is unimportant for the regular ISL procedure. The probe can be used in the isothermal signal amplification procedure if it contains a single recognition site for a blunt-end restrictase (Sma I in this example). (**b**) In the regular ISL, probes are ligated to 5′PO_4 blunt-ended apoptotic DNA breaks in tissue sections. They cannot ligate to each other due to the absence of 5′PO_4 termini required by T4 DNA ligase. Therefore, the regular ISL reaction stops when all DNA probes are ligated to 5′PO_4 breaks in apoptotic DNA

tissue section processing, cutting, or heating. In addition, 3′OH termini in DNA are generated in necrosis, nonlethal DNA damage, and DNA repair. All of these can produce TUNEL staining not related to apoptosis. This shortcoming of the TUNEL assay is discussed in a number of cautionary notes [6, 9–17]. In spite of this limitation TUNEL is far more popular than the more specific ISL technique. One of the reasons is the time-consuming nature of ISL, requiring 18 h-long incubations, much more protracted than the ~3 h-long TUNEL [18].

Accelerating ISL detection would make it significantly more attractive for routine apoptosis labeling in the tissue section format. However, the considerable length of incubation in ISL is essential for the ligation of a sufficient number of probes to DNA breaks in tissue sections. The probe attachment to DNA ends that are hidden inside tissue section is slower than DNA ligations occurring entirely in solution. This is probably due to the lower accessibility of the cellular ligation targets which are partially concealed in the cross-linked matrix of the section. The ISL labeling reaction is not accelerated by commercially available ultrafast ligation mixes exploiting the volume exclusion effect of PEG, because similar or even higher PEG concentrations are already used in ISL protocols [5, 19, 20]. The same is true for boosting the T4 DNA ligase concentration, as the standard ISL protocol already uses the highly concentrated enzyme [5, 19]. Besides, following this approach even further can make the assay prohibitively expensive.

The other way to make this assay faster is by a major increase of the probe signal. Interestingly, such an approach was employed in the biochemical counterpart of ISL [2] which appeared soon after we developed the in situ ligation method [6]. Then a paper was published in 1997 describing a PCR-based technique, which also used ligase, this time for biochemical detection of double-strand breaks in DNA of apoptotic cells [2]. In this study, nucleosomal ladders were revealed by a ligation-mediated polymerase chain reaction (LMPCR), which amplified DNA fragments with blunt, 5′ phosphorylated ends. LMPCR combined detection and PCR-based amplification steps into a single procedure and was therefore fast and sensitive.

However, the straightforward adaptation of PCR-based amplification to the ISL technique proved to be difficult. Therefore, we developed an alternative amplification technique, which specifically enhances ISL signal. This new amplification procedure eliminates the need for the time-consuming incubations with ligase because sufficient ISL signal is achieved after 1 h incubation.

The ISL signal is enhanced with the help of the new isothermal cyclic elongation reaction that makes the already ligated probes much longer by generating their protracted labeled extensions—"zebra tails." These are produced by repetitive ligations of the new probes to the continually reactivated ends of the growing "seed" probes. The term "zebra tail" originates from the graphical representation of this perpetual lengthening when each new ligation event is marked with a black stripe. The ongoing probe extension produces the familiar "zebra" pattern.

Remarkably, the "zebra tail" signal amplification is initiated by introducing a very minimal change to the standard ISL protocol. The amplification cycles start after the addition of a restriction enzyme to the labeling mix. The added restrictase is required to be a blunt-end cutter and to have a single cleavage site in the PCR fragment used as an ISL probe. The principles of the zebra tail amplification (ZTA) technique are illustrated in Fig. 2.

In sum, the amplification reaction constantly regenerates the ligatable blunt end of the growing probe enabling its continuous elongation. Every cleavage-ligation cycle adds either the full- or the partial-length of the initial labeled DNA fragment (*see* **Note 1**). When each new ligation event is marked with a black stripe, the multiple cycles of cleavage-ligation produce a long "zebra tail" of labeled DNA (Fig. 2) (*see* **Note 2**).

Because "zebra tails" are the direct extensions of the reacted probes, they do not disperse or diffuse and indicate the probe's location with high accuracy. This property can be useful in providing better resolution during in situ detection.

The presented protocol is applicable for specific non-fluorescent detection of apoptotic cells in tissue sections. The procedure is eightfold faster compared to the original ISL and overall takes ~3 h to complete.

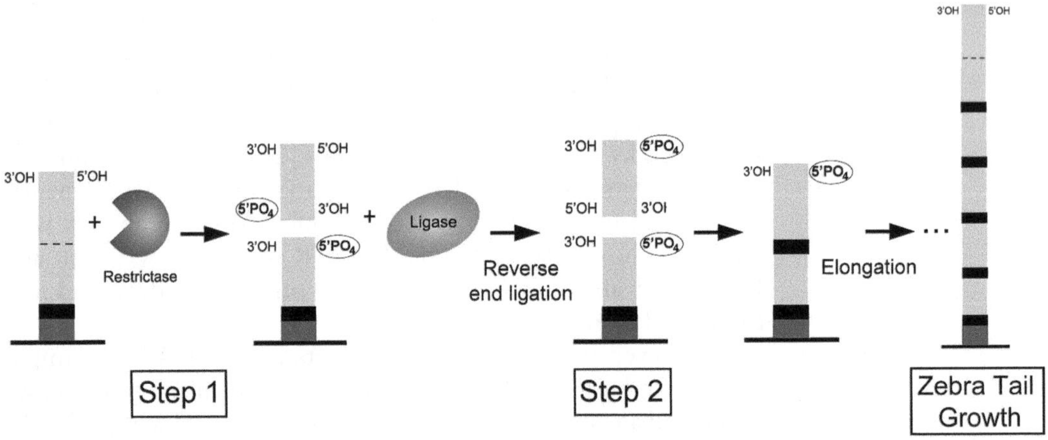

Fig. 2 Zebra tail amplification of ISL signal. **Step 1.** A restrictase present in the labeling mix cleaves the ligated ISL probes in the section and produces the new generation of blunt ends with $5'PO_4$. These serve as selective targets for the second round of probe ligation. **Step 2.** Any ISL probe in solution, regardless if it is cut or not (has $5'PO_4$ or $5'OH$ at the ends), can now ligate to the cleaved probes in the section possessing $5'PO_4$ at the ends. In some cases it can restore the original probe (when $3'OH/5'PO_4$ end of the solute probe is ligated to $3'OH/5'PO_4$ end of the probe in the section); this will be immediately re-cleaved and will not hamper the amplification reaction (not shown). In other cases, such as the reverse end ligation, (when $3'OH/5'OH$ end of the solute probe is ligated to $3'OH/5'PO_4$ end of the probe in the section) the restriction site is not regenerated, so the extended probe cannot be re-cut. This permanently elongates the "seeded" probe and exposes the new $5'PO_4$ at its growing end, thus enabling another round of extension. The newly-generated $5'PO_4$ at blunt ends are responsible for the growth of the labeled DNA strand and the ISL signal amplification. "Zebra tail" image is formed when each new ligation event on the growing DNA strand is marked with a black stripe

In addition to the time economy, the increased signal intensity can be helpful for better visualization of morphological changes in individual apoptotic cells when strong and selective labeling of the apoptotic nuclear chromatin is needed (*see* **Note 3**). The assay can also be applied for specific identification of minute apoptotic bodies in preparations of microvesicles, exosomes, and other extracellular vesicles [21].

2 Materials

1. The 5–6 μm-thick sections cut from paraformaldehyde-fixed, paraffin-embedded tissue blocks onto charged and precleaned slides.
2. Xylene.
3. 70, 80 and 96% Ethanol.
4. T4 DNA ligase 5 U/μL (Roche Molecular Biochemicals, Indianapolis, IN) (*see* **Note 4**).

5. 10× reaction buffer for T4 DNA ligase: 660 mM-Tris–HCl, 50 mM $MgCl_2$, 10 mM dithioerythritol, 10 mM ATP, pH 7.5 (20 °C) (Roche Molecular Biochemicals, Indianapolis, IN).
6. 30% (w/v) solution of PEG-8000 (Sigma, St. Louis, MO) in bidistilled water (*see* **Note 5**).
7. Proteinase K (Roche Molecular Biochemicals, Indianapolis, IN) 20 mg/mL stock solution in distilled water. Store at −20 °C. In the reaction use 50 μg/mL solution in PBS, prepared from the stock. Do not reuse (*see* **Note 6**).
8. PCR-derived digoxigenin-labeled DNA fragment obtained using Pfu polymerase (*see* **Note 7**).

   The sequence of the fragment is not important. However, it should contain a unique restriction site. We used the same 226 bp long PCR-labeled fragment as before in regular ISL [6, 19]. It contained a single Sma I restriction site (*see* **Note 8**).
9. Sma I restriction endonuclease 20 U/μL (New England Biolabs, Indianapolis, IN). This enzyme is used only with the PCR fragment described above. If another fragment is used, check its restriction map, and select the appropriate blunt end-producing restrictase (*see* **Note 9**).
10. Phosphate-buffered saline (PBS), pH 7.4.
11. Washing buffer: 100 mM Tris–HCl, pH 7.5; 100 mM NaCl.
12. Blocking solution: 10% sheep serum in 100 mM Tris–HCl, pH 7.5; 100 mM NaCl.
13. 1-Step™ NBT/BCIP (Pierce)—ready-to-use substrate for alkaline phosphatase that yields a very intense purple signal. If background caused by endogenous phosphatases is a problem, then 1-Step NBT/BCIP plus Suppressor can be used. This substrate formulation additionally contains 1 mM levamisole to inhibit endogenous phosphatase activity.
14. Anti-digoxigenin antibody—alkaline-phosphatase conjugate (Roche Molecular Biochemicals, Indianapolis, IN). Fab fragments from an anti-digoxigenin antibody from sheep, conjugated with alkaline phosphatase for the detection of digoxigenin-labeled compounds.
15. 22 × 22 mm or 22 × 40 mm glass or plastic coverslips. Plastic coverslips are preferable as they are easier to remove from the section.
16. Aqueous slide mounting media such as Aqua-Mount (Thermo Fisher Scientific, Waltham, MA) or ImmunoHistoMount (Santa Cruz Biotechnology, Santa Cruz, CA).
17. Microscope with appropriate filters and objectives.

3 Method

1. Place the sections in a slide rack and dewax in xylene for 5 min, transfer to a fresh xylene bath two more times for an additional 2 min each. For each transfer, dip cassette up and down three times.
2. Rehydrate by passing through graded ethanol concentrations: 96% Ethanol—2 × 2 min; 80% Ethanol—2 × 2 min; 70% Ethanol—2 min; water—2 × 2 min. For each transfer, dip cassette up and down three times.
3. Digest section with Proteinase K. Use 100 μL of a 50 μg/mL solution per section. Incubate for 10 min at room temperature in a humidified chamber (*see* **Note 10**).
4. Rinse in distilled water for 3 × 2 min. For each transfer, dip cassette up and down three times.
5. Apply 100 μL per section of the pre-incubation solution. Incubate for 5 min at room temperature (23 °C). The pre-incubation solution consists of a 1× T4 DNA ligase reaction buffer supplemented with PEG-8000 to the final concentration of 15%. It contains 66 mM-Tris HCl, pH 7.5, 5 mM $MgCl_2$, 1 mM dithioerythritol, 1 mM ATP, and 15% polyethylene glycol (*see* **Note 11**).
6. Aspirate the pre-incubation solution and apply the full ligase reaction mix with the probe (750 ng/μL stock) and T4 DNA ligase (*see* **Note 12**).

   In situ ligation labeling solution (20 μL per section):
   Prepare on ice in this order:
   4 μL—bidistilled water.
   10 μL—PEG8000 (30% stock solution).
   2 μL—10× buffer for T4 DNA ligase.
   1 μL—PCR labeled probe (0.5 μg/μL).
   2 μL—T4 DNA ligase (5 U/μL) (*see* **Note 4).**
   1 μL—Sma I restrictase (20 U/μL) (*see* **Note 9**).

   The total volume of the labeling solution can be scaled up to accommodate the bigger sections.

   Incubate for 1 h. at room temperature (23 °C) (*see* **Note 13**) in a humidified chamber with a plastic coverslip.

   The total volume of the labeling solution can be scaled up to accommodate the bigger sections.
7. Remove coverslips by gently immersing the slides vertically in a Coplin jar containing water at room temperature. Then wash sections 3 × 5 min in distilled water.
8. Apply 100 μL per section of the blocking solution containing 10% sheep serum. Incubate for 15 min at room temperature (23 °C) in a humidified chamber.

9. Aspirate the pre-incubation solution and apply 50 μL per section of solution of anti-digoxigenin antibody—alkaline-phosphatase conjugate diluted 1:100 in the blocking solution. Incubate for 10 min at room temperature (23 °C) in a humidified chamber.
10. Wash 2 × 10 min in 100 mM Tris–HCl, pH 7.5; 100 mM NaCl.
11. Apply 1-Step™ NBT/BCIP substrate solution for alkaline phosphatase. Keep in the dark. Monitor color development. The color usually develops in 5–20 min.
12. Stop reaction by rinsing sections in water. Add water-soluble mounting media such as Aqua-Mount or ImmunoHistoMount and coverslip.

4 Notes

1. The elongation reaction is a single-strand ligation occurring between 3′OH and 5′PO_4 termini, whereas 3′OH and 5′OH termini cannot be linked by T4DNA ligase. Therefore, each fully formed double-strand "zebra tail" contains the regularly positioned nicks in one opposing strand.
2. Some free-floating "zebra tails" are also created in solution, where restriction and amplification can occur prior to probe ligation to DNA breaks in the section. This does not change the results of the labeling reaction, because the longer probes created in this case have the same configuration of ends as the original fragment. When ligated to breaks in the section, they detect the same type of DNA damage and produce enhanced signal.
3. Both ISL and its ZTA enhancement presented here can selectively detect blunt-ended and 5′ phosphorylated DNA breaks. This is convenient for apoptosis labeling because the majority of DNA breaks in apoptosis are blunt-ended with the 3′OH/5′PO_4 at the ends [1–3]. However, smaller numbers of these same breaks are produced in other important situations, such as V(D)J recombination [22], as a result of action of ionizing radiation, various chemicals, free radicals, and biological processes [23, 24].
4. In our experiments the highly concentrated (5 U/μL) (Roche Molecular Biochemicals, Indianapolis, IN) ligase preparation repeatedly gives the best signal.
5. DNA ligation is strongly stimulated by 15% PEG-8000 in the reaction solution due to the volume exclusion effect [20]. However, sometimes PEG presence can cause nonspecific background staining. In such cases reducing PEG concentration can significantly reduce or eliminate the problem.

6. In high concentrations (1 mg/mL and higher) proteinase K is a very stable enzyme and can be stored for extended time. However, it gradually loses activity when stored in aqueous solutions at low concentrations (~10 μg/mL) [25].
7. Pfu is a proofreading polymerase and exclusively produces blunt-ended DNA fragments in PCR. The non-proofreading polymerases, such as Taq, produce PCR fragments, which contain single base extensions at their 3′ ends. This prevents ZTA from progression because the fragments with single base overhangs do not ligate to the blunt ends of the growing "zebra tails."

   In PCR probe labeling we prefer using digoxigenin over biotin as it produces less background staining in our experiments. In general, Pfu efficiently incorporates digoxigenin-11-dUTP or biotin -dUTP if the ratio of substituted-dUTP to dTTP in the labeling mix is 1:2. Therefore usually ~35% of a dTTP in the Pfu PCR reaction mix is substituted by its tagged analog [19]. This is expected to translate into the substitution of ~35% of thymidines in our 226-bp probe by the tagged nucleotides and would result in the insertion of 26 tags. The PCR labeling of in situ ligation probes and their application is described in detail elsewhere [19].
8. The complete sequence of the fragment is presented in [26]. However, this specific sequence has little importance for the labeling reaction. Almost any 100–500-bp long PCR-derived fragment can be used. This range of probe lengths was successfully employed by different laboratories for the original ISL procedure [6, 27–35], and can serve as a starting point for ZTA, which is an accelerated ISL staining. Much longer probes are impractical as they can break during purification and storage, producing smaller fragments with nonspecific configurations of ends.
9. Different restrictases can be used to drive the reaction. Although we have successfully used frequent cutters, such as Alu I, and more rare cutters such as Sma I, but it is probably better to use rare cutters whenever possible, because the fewer restriction sites per genome potentially limit the likelihood of creating artifactual blunt-ended DNA breaks in the section. In any case the control sections are still needed. Such control sections with confirmed absence of ds DNA breaks should be treated with restriction enzyme alone and then tested by in situ ligation to exclude the possibility of the creation of additional DNA breaks. In our experience, short time (1 h) incubation with Sma I did not create DNA breaks in formalin-fixed human sections, most likely because DNA in the formalin-fixed tissues is cross-linked and is not readily available for cleavage by restrictases. Sma I is used at room temperature (25 °C is recommended by New England Biolabs). At 37 °C it has a half-life of only 15 min.

10. Proteinase K digestion time may need adjustment depending on the tissue type. Insufficient digestion may result in a weaker signal. Overdigestion results in signal disappearance and section disruption. Most often 10–15 min digestion is sufficient.
11. Pre-incubation with ligation buffer ensures even saturation of the section prior to the addition of the enzyme and the probe, and was shown to increase the ligation efficiency. Prepare pre-incubation solution by adding 10 μL of the 10× ligase buffer (Roche) to 40 μL of bidistilled water and 50 μL of 30% PEG-8000.
12. A mock reaction is recommended as a regular control in order to rule out nonspecific background staining. In the mock reaction an equal volume of 50% glycerol in water is substituted for T4 DNA ligase.
13. Lowering of the temperature to 16 °C reduces the signal; a temperature increase to 37 °C completely eliminates the signal.

Acknowledgment

I am grateful to Candace Minchew for her outstanding technical assistance.

This research was supported by grant R01 NS082553 from the National Institute of Neurological Disorders and Stroke, National Institutes of Health and by grants R21 CA178965 from the National Cancer Institute, National Institutes of Health and R21 AR066931 National Institute of Arthritis and Musculoskeletal and Skin Diseases, National Institutes of Health.

References

1. Nagata S, Nagase H, Kawane K et al (2003) Degradation of chromosomal DNA during apoptosis. Cell Death Differ 10:108–116
2. Staley K, Blaschke A, Chun J (1997) Apoptotic DNA fragmentation is detected by a semiquantitative ligation-mediated PCR of blunt DNA ends. Cell Death Differ 4:66–75
3. Didenko VV, Ngo H, Baskin DS (2003) Early necrotic DNA degradation: presence of blunt-ended DNA breaks, 3′ and 5′ overhangs in apoptosis, but only 5′ overhangs in early necrosis. Am J Pathol 162:1571–1578
4. Hornsby PJ, Didenko VV (2011) In situ ligation: a decade and a half of experience. Methods Mol Biol 682:49–63. doi:10.1007/978-1-60,327-409-8_5
5. Didenko VV (2002) Detection of specific double-strand DNA breaks and apoptosis in situ using T4 DNA ligase. Methods Mol Biol 203:143–151
6. Didenko VV, Hornsby PJ (1996) Presence of double-strand breaks with single-base 3′ overhangs in cells undergoing apoptosis but not necrosis. J Cell Biol 135:1369–1376
7. Gavrieli Y, Sherman Y, Ben-Sasson SA (1992) Identification of programmed cell death in situ via specific labeling of nuclear DNA fragmentation. J Cell Biol 119:493–501
8. Loo DT (2011) In situ detection of apoptosis by the TUNEL assay: an overview of techniques. Methods Mol Biol 682:3–13. doi:10.1007/978-1-60,327-409-8_1
9. Charriaut-Marlangue C, Ben-Ari Y (1995) A cautionary note on the use of the TUNEL stain to determine apoptosis. Neuroreport 7:61–64

10. Wolvekamp MC, Darby IA, Fuller PJ (1998) Cautionary note on the use of end-labeling DNA fragments for detection of apoptosis. Pathology 30:267–271
11. Grasl-Kraupp B, Ruttkay-Nedecky B, Koudelka H et al (1995) In situ detection of fragmented DNA (TUNEL assay) fails to discriminate among apoptosis, necrosis, and autolytic cell death: a cautionary note. Hepatology 21:1465–1468
12. Sloop GD, Roa JC, Delgado AG et al (1999) Histologic sectioning produces TUNEL reactivity. A potential cause of false-positive staining. Arch Pathol Lab Med 123:529–532
13. Pulkkanen KJ, Laukkanen MO, Naarala J, Yla-Herttuala S (2000) False-positive apoptosis signal in mouse kidney and liver detected with TUNEL assay. Apoptosis 5:329–333
14. Bassotti G, Villanacci V, Fisogni S et al (2007) Comparison of three methods to assess enteric neuronal apoptosis in patients with slow transit constipation. Apoptosis 12:329–332
15. Lawrence MD, Blyth BJ, Ormsby RJ et al (2013) False-positive TUNEL staining observed in SV40 based transgenic murine prostate cancer models. Transgenic Res 22:1037–1047. doi:10.1007/s11248-013-9694-7
16. Haunstetter A, Izumo S (2012) Strategies to prevent apoptosis. In: Hasenfuss G, Marban E (eds) Molecular approaches to heart failure therapy, 3rd edn. Springer Science & Business Media, Berlin
17. Galluzzi L, Aaronson SA, Abrams J et al (2009) Guidelines for the use and interpretation of assays for monitoring cell death in higher eukaryotes. Cell Death Differ 16:1093–1107
18. Apoptosis Detection Using Terminal Transferase and Biotin-16-dUTP (TUNEL Enzyme Method). (2016) http://www.ihcworld.com/_protocols/apoptosis/tunel_enzyme.htm.
19. Didenko VV (2011) In situ ligation simplified: using PCR fragments for detection of double-strand DNA breaks in tissue sections. Methods Mol Biol 682:65–75. doi:10.1007/978-1-60,327-409-8_6
20. Maunders MJ (1993) DNA and RNA ligases (EC 6.5.1.1, EC 6.5.1.2, EC 6.5.1.3). Methods Mol Biol 16:213–230. doi:10.1385/0-89,603-234-5:213
21. Hauser P, Wang S, Didenko VV (2017) Apoptotic bodies: selective detection in extracellular vesicles. Methods Mol Biol 1554:200–193. doi:10.1007/978-1-4939-6759-9_12
22. Schlissel M, Constantinescu A, Morrow T et al (1993) Double-strand signal sequence breaks in *V(D)J* recombination are blunt, 5′-phosphorylated, RAG-dependent, and cell cycle regulated. Genes Dev 7:2520–2532
23. van Gent DC, Hoeijmakers JHJ, Kanaar R (2001) Chromosomal stability and the DNA double-stranded break connection. Nat Rev Genet 2:196–206
24. Longhese MP, Guerini I, Baldo V, Clerici M (2008) Surveillance mechanisms monitoring chromosome breaks during mitosis and meiosis. DNA Repair 7:545–557
25. Sweeney PJ, Walker JM (1993) Proteinase K (EC 3.4.21.14). Methods Mol Biol 16:305–311. doi:10.1385/0-89,603-234-5:305
26. Yang L, Didenko VV, Noda A et al (1995) Increased expression of p21Sdi1 in adrenocortical cells when they are placed in culture. Exp Cell Res 221:126–131
27. Koda M, Takemura G, Kanoh M et al (2003) Myocytes positive for in situ markers for DNA breaks in human hearts which are hypertrophic, but neither failed nor dilated: a manifestation of cardiac hypertrophy rather than failure. J Pathol 199:229–236
28. Okada H, Takemura G, Koda M et al (2005) Myocardial apoptotic index based on in situ DNA nick end-labeling of endomyocardial biopsies does not predict prognosis of dilated cardiomyopathy. Chest 128:1060–1062
29. Schoppet M, Al-Fakhri N, Franke FE et al (2004) Localization of osteoprotegerin, tumor necrosis factor-related apoptosis-inducing ligand, and receptor activator of nuclear factor-kappa B ligand in Mönckeberg's sclerosis and atherosclerosis. J Clin Endocrinol Metab 89:4104–4112
30. Audo I, Darjatmoko SR, Schlamp CL et al (2003) Vitamin D analogues increase p53, p21, and apoptosis in a xenograft model of human retinoblastoma. Invest Ophthalmol Vis Sci 44:4192–4199
31. Al-Fakhri N, Chavakis T, Schmidt-Woll T et al (2003) Induction of apoptosis in vascular cells by plasminogen activator inhibitor-1 and high molecular weight kininogen correlates with their anti-adhesive properties. J Biol Chem 384:423–435
32. Matsuoka R, Ogawa K, Yaoita H et al (2002) Characteristics of death of neonatal rat cardiomyocytes following hypoxia or hypoxia-reoxygenation: the association of apoptosis and cell membrane disintegrity. Heart Vessels 16:241–248
33. Guerra S, Leri A, Wang X et al (1999) Myocyte death in the failing human heart is gender dependent. Circ Res 85:856–866
34. Leri A, Claudio PP, Li Q et al (1998) Strech-mediated release of angiotensin II

induces myocyte apoptosis by activating p53 that enhances the local renin-angiotensin system and decreases the Bcl-2 to Bax protein ratio in the cell. J Clin Invest 101:1326–1342

35. Murata I, Takemura G, Asano K et al (2002) Apoptotic cell loss following cell proliferation in renal glomeruli of Otsuka Long-Evans Tokushima Fatty rats, a model of human type 2 diabetes. Am J Nephrol 22:587–595

Part III

Accelerated Detection (Time-Saving Versions of Conventional Techniques)

Chapter 16

Twelve-Gel Comet Assay Format for Quick Examination of DNA Damage and Repair

Sergey Shaposhnikov and Andrew Collins

Abstract

The comet assay (single cell gel electrophoresis) is a sensitive, versatile method for detecting DNA damage in eukaryotic cells. The traditional comet assay format has 1 or 2 gels on a microscope slide, 1 sample per slide, and there is a limit of 40 gels per experiment given the size of a typical electrophoresis tank. To increase throughput, we have designed and tested a system with 12 minigels on one slide, allowing analysis of up to 12 times more samples in one electrophoresis run. The novel comet assay format compares well with the traditional technology. The various steps are suitable for further automation, and the formats can be adapted to fully automated scoring. The new procedures save time at all stages as fewer slides are handled, and the amounts of reagents needed are reduced significantly. This format is particularly useful for testing of numerous genotoxic agents and nanomaterials at different concentrations and on different types of cells; simultaneous analysis of different lesions using a range of enzymes; and analysis of cell extracts for DNA repair activity.

Key words Comet assay, High-throughput genotoxicity assay, DNA damage, DNA repair

1 Introduction

The comet assay is the method of choice for measuring DNA damage in cellular DNA [1]. Briefly, cells in suspension are embedded in a thin layer of agarose on a microscope slide, lysed, and electrophoresed.

The lysis step includes incubation in a solution containing Triton X-100, which breaks membranes, and high salt, which strips off histones. This leaves the DNA in a nucleosome-free form, which remains attached to the nuclear matrix in a structure known as a nucleoid.

On electrophoresis, DNA is attracted toward the anode, but migration occurs only if breaks are present. The migrating DNA from each cell forms a comet-like structure when viewed by fluorescence microscopy, and the fraction of DNA in the tail is proportional to the frequency of breaks.

Vladimir V. Didenko (ed.), *Fast Detection of DNA Damage: Methods and Protocols*, Methods in Molecular Biology, vol. 1644, DOI 10.1007/978-1-4939-7187-9_16, © Springer Science+Business Media LLC 2017

The assay can be combined with lesion-specific endonucleases, so that damage other than simple strand breaks can be measured. Formamidopyrimidine DNA glycosylase (Fpg) is used to detect oxidized purines (mainly 8-oxoGua), endonuclease III for oxidized pyrimidines, and T4 endonuclease V for UV-induced cyclobutane pyrimidine dimers.

Repair of different kinds of DNA damage can be monitored simply by incubating cells after treating them with a specific damaging agent, and measuring the damage remaining (e.g., strand breaks, or enzyme-sensitive sites) at intervals during the incubation. Alternatively, a cell extract is incubated with a substrate of nucleoids containing specific DNA lesions, and the capacity of enzymes in the extract to introduce DNA breaks is a measure of repair activity.

The advantages of the assay are its simplicity, speed, sensitivity, applicability to different cell types, avoidance of radioactive labeling, and—perhaps most significantly—the fact that damage is assessed at the level of individual cells.

The limitation of the traditional version of the assay is that it is not suitable for large screening studies, in which a high-capacity robust and standardized assay is needed. The conventional comet assay format has 1 or 2 gels on a microscope slide. Processing of samples is time consuming, and there is a limit of 40 gels per experiment, given the size of a typical electrophoresis tank.

To optimize the method, we have developed an alternative 12-gel comet assay format [2]: Using a simple template, 12 minigels are set in a 6 × 2 array on a microscope slide.

If the gels are to be treated with different reagents or enzymes, the slide is clamped in a chamber device and the gels are separated by a gasket placed on the slide with the holes positioned over the gels so that incubations can be carried out independently (Fig. 1).

With the use of the novel 12 gel format several times more (up to 240) samples can be analyzed with one electrophoresis run, and fewer cells and smaller volumes of test solutions are required.

Applications of the modified method include treatment with genotoxic agents at different concentrations, simultaneous analysis of different lesions using a range of enzymes, and analysis of cell extracts for DNA repair activity.

The novel comet assay format compares well with the traditional technology, with the advantage of higher throughput. The various steps are suitable for further automation, and the formats can be adapted to fully automated scoring. The new procedures save time at all stages as fewer slides are handled and reduce the amounts of reagents needed.

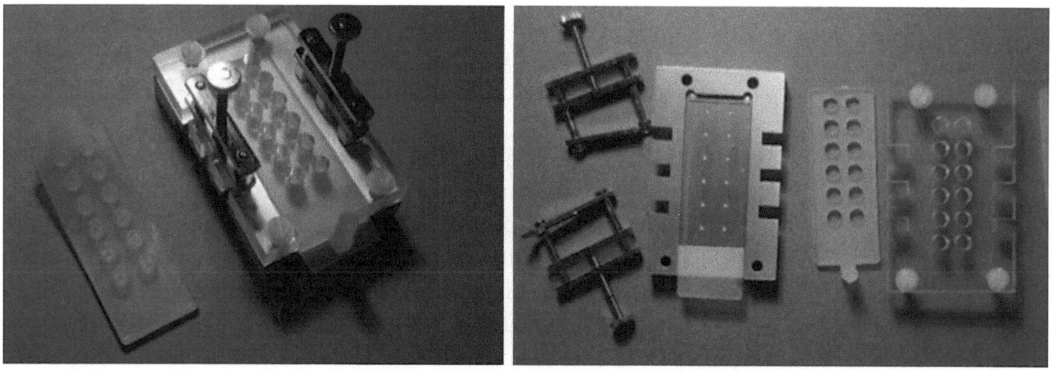

Fig. 1 Component parts of a 12-gel chamber unit, including metal base with marks for positioning gels on slide, silicone rubber gasket, plastic top-plate with wells, and silicone rubber seal

2 Materials

2.1 Equipment and Supplies

1. Staining jars (vertical or horizontal).
2. Glass slides (frosted end).
3. Parafilm.
4. Incubator for 37 °C.
5. Moist chamber (e.g., glass or plastic box with platform for slides above a layer of water).
6. Eletrophoresis tank (horizontal).
7. Electrophoresis power supply.
8. Fluorescence microscope.
9. Incubation chamber with silicon gasket (Norgenotech AS, Norway).

2.2 Reagents and Solutions

1. Normal melting point agarose (NMP agarose): 1% in H_2O. A few 100 mL is enough for several hundred slides.
2. Low melting point agarose (LMP agarose): 1% in PBS. Store at 4 °C in 10 mL aliquots.
3. Phosphate-buffered saline (PBS).
4. Reagents for cell culture.
5. Histopaque 1077 (Sigma) or Lymphoprep (Axis-Shield AS, Norway).
6. Lysis solution: 2.5 M NaCl, 0.1 M EDTA, 10 mM Tris. Prepare 1 L. Set pH to 10 with 10 M NaOH solution. (Add 35 mL of NaOH straight away to dissolve EDTA, and then add dropwise to pH 10). Add 1 mL Triton X-100 per 100 mL immediately before use.
7. Alkaline electrophoresis solution: 0.1 M NaOH, 1 mM EDTA.

8. Enzyme reaction buffer: 40 mM HEPES, 0.1 M KCl, 0.5 mM EDTA, 0.2 mg/mL BSA, pH 8.0 with KOH (can be made as 10 × stock, adjusted to pH 8.0 and frozen at −20 °C).

2.3 Enzymes

Endonuclease III (endo III), formamidopyrimidine glycosylase (Fpg), and T4 endonuclease V are commercially available in purified form, or may be obtained from a laboratory producing them. They are isolated from bacteria containing over-producing plasmids. Because such a high proportion of protein is the enzyme, a crude extract is satisfactory, as nonspecific nuclease activity is not significant at the concentrations employed. The final dilution of the working solution will vary from batch to batch. Follow supplier's instructions, or use the following as a guide (assuming a final dilution of 3000×) (*see* **Note 1**).

FPG:

1. On receipt, dispense the stock solution into 5 μL aliquots and refreeze at −80 °C. This is to minimize repeated freezing and thawing.
2. Take one of these aliquots and dilute to 0.5 mL using the enzyme reaction buffer—with the addition of 10% glycerol. Dispense into 10 μL aliquots (label as "100× diluted") and freeze at −80 °C.
3. For use, dilute one of these 10 μL aliquots to 300 μL with buffer (no glycerol) and keep on ice until you add it to the gels: do not refreeze this working solution.

Endo III and T4 endonuclease V:

1. Dispense the stock solution into 5 μL aliquots and refreeze at −80 °C.
2. Take one of these aliquots and dilute to 0.5 mL using the enzyme reaction buffer (no need to add glycerol as Endo III is more stable). Dispense this into 10 μL aliquots (label as "100× diluted") and freeze at −80 °C.
3. For use, dilute one of these 10 μL aliquots to 300 μL with buffer (no glycerol) and keep on ice until you add it to the gels.

3 Methods

3.1 Slide Preparation for the Comet Assay (Precoating)

1. The slides for precoating should be grease-free; clean if necessary by soaking in alcohol and then wiping dry with a clean tissue.
2. Dip slides in a vertical staining jar of melted 1% standard agarose in H_2O.

3. Drain off excess agarose, wipe the back clean, and dry by leaving on a clean bench overnight of attached cells in culture or after disaggregation of tissue at 2.5×10^5 cells/mL in PBS.
4. Quickly add 140 μL of 1% LMP agarose to 30 μL of cells at 37 °C and mix by aspirating agarose up and down (once) with pipettor.
5. Place twelve 5 μL drops onto the precoated slide as two rows of six (without coverslips). Allow gels to set at 4 °C for 5 min.

3.2 Lysis

1. Add 1 mL Triton X-100 to 100 mL of lysis solution (4 °C).
2. Place in this solution in a (horizontal) staining jar.
3. Leave at 4 °C for a period of at least 1 h (overnight lysis is possible).

3.3 Gels Treatment with Enzymes or Reagents

1. (For incubation with different reagents proceed from **step 2** below), for enzyme treatment wash slides in three changes of the enzyme buffer at 4 °C in a staining jar, for 5 min each; remove slides from last wash, and dab off excess liquid with tissue.
2. Place the gasket on the slide with the holes positioned over the gels, and held under light pressure in a specially designed chamber (*see* Fig. 1).
3. Pipette reagents or enzymes onto the gels, and place the chamber in a humid box at 37 °C for 30 min (or use other conditions suitable for your purposes).

3.4 Alkaline Electrophoresis

1. Gently place slides on platform in tank, immersed in alkaline electrophoresis solution (*see* **Note 2**), forming complete rows (fill gaps with blank slides).
2. Make sure that tank is level and gels are just covered. Leave for 20 min.
3. For most tanks (i.e., of standard size), run at 25 V (constant voltage setting) for 20 min.
4. If there is too much electrolyte covering the slides, the current may be so high that it exceeds the maximum—so set this at a higher level than you expect to need. If necessary, i.e., if 25 V is not reached, remove some solution. Normally, the current is around 300 mA but this is not crucial (*see* **Note 3**).
5. Neutralize by washing for 10 min with PBS in a staining jar at 4 °C, followed by 10 min wash in water.
6. Proceed to staining while gels are wet.
7. Dry (room temperature) for microscopy or storage.

3.5 Staining and Scoring

Use appropriate DNA stain according to the recommendation of the supplier. We normally stain in SYBR Gold diluted as 1:10,000 in TE-buffer for 20 min. Depending on the number of slides staining can be done either under coverslips or in jars. After staining the preparations are examined using a fluorescence microscope coupled with a software designed for scoring comets. We use Comet IV from Perceptives (UK) and Pathfinder from Imstar (France).

4 Notes

1. The buffer in which enzyme is stored may contain β-mercaptoethanol to preserve the enzyme. However, inclusion of sulfhydryl reagents in the reaction buffer would significantly increase background DNA breakage.
2. Electrophoresis solution should be cooled before use, e.g., by pouring into the electrophoresis tank in the cold room an hour or so before it is needed.
3. The voltage depends on tank dimensions. 0.8 V/cm is recommended, calculated on the basis of the distance across the platform (where the conducting layer is least deep and the resistance highest). The changes in voltage/current/resistance across the tank from electrode to electrode, and the conditions within the gel, provide an interesting exercise in simple theoretical physics.

Acknowledgment

This project was partially financed by the EU FP7 project NANoREG, grant 310584, and Research Council of Norway funded French-Norwegian project ComPack, grant 13-03 FNS.

References

1. Collins AR (2004) The comet assay for DNA damage and repair. Mol Biotechnol 26:249–261
2. Shaposhnikov S, Azqueta A, Henriksson S, Meier S, Gaivão I, Brunborg G, Nilsson M, Collins AR (2010) Twelve-gel slide format optimised for comet assay and fluorescent in situ hybridisation. Tox Letters 19(195): 31–34

Chapter 17

Immunofluorescence Analysis of γ-H2AX Foci in Mammalian Fibroblasts at Different Phases of the Cell Cycle

Liudmila Solovjeva, Denis Firsanov, Nadezhda Pleskach, and Maria Svetlova

Abstract

H2AX phosphorylation at Ser139 (formation of γ-H2AX) is an indicator of double-strand breaks in DNA (DSBs) after the action of different genotoxic stresses, including ionizing radiation, environmental agents, and chemotherapy drugs. The sites of DSBs can be visualized as focal sites of γ-H2AX using antibodies and immunofluorescence microscopy. The microscopy technique is the most sensitive method of DSB detection in individual cells. It is useful for experimental research, radiation biodosimetry, and clinical practice. In this chapter, we provide an immunochemical protocol for γ-H2AX labeling and analysis by confocal microscopy. The advantage of the assay is that it enables the quantitation of γ-H2AX foci in individual cells in different phases of the cell cycle.

Key words Immunofluorescence, Confocal microscope, γ-H2AX, Double-strand breaks, Apoptosis, Mammalian cells

1 Introduction

Phosphorylation of H2AX is an early event in DNA damage response. After DSB induction, histone H2AX is rapidly phosphorylated at Ser139 (forming γ-H2AX) at the ends of the breaks, and serves for accumulation of DSB repair proteins [1]. γ-H2AX can be visualized as distinct foci in the nuclei of irradiated cells. It has been shown that each γ-H2AX focus in the nucleus corresponds to one DSB induced by ionizing radiation [2]. The formation of γ-H2AX is not restricted to DSB signaling only, but is associated with some other cellular processes that require chromatin remodeling. For example, H2AX is phosphorylated during preimplantation development of oocytes. It is required for X chromosome inactivation in female somatic cell and mitosis [3].

Vladimir V. Didenko (ed.), *Fast Detection of DNA Damage: Methods and Protocols*, Methods in Molecular Biology, vol. 1644, DOI 10.1007/978-1-4939-7187-9_17,

DSBs are induced by free radical mechanism after the action of ionizing radiation (IR) or the action of some radiomimetic drugs: bleomycin, calicheamicin, neocarcinostatin, and others [4]. Replication-mediated DSBs at the sites of stalled replication forks are induced by topoisomerase I inhibitor camptothecin [5], UV lesions, alkylated bases [6, 7], nucleoside analogs, and other drugs used in chemotherapy [8].

The phosphorylated form of H2AX can be detected in cell nuclei with antibodies specific for γ-H2AX and visualized by different techniques: immunoblotting, flow cytometry, and immunofluorescence microscopy. Among these methods, immunofluorescence microscopy is the most sensitive. Quantification of γ-H2AX foci in X-ray-irradiated cells is possible even after exposure to doses less than 1 mGy [1]. Immunofluorescence microscopy-based γ-H2AX detection is used in different research area including detection of DSBs after IR, after the action of different cytotoxic chemical agents and environmental damage, in the study of DSB repair kinetics, and aging research. This method is applied also in biodosimetry, in the study of chronic inflammation, and in cancer treatment.

Flow cytometry is the method of choice for the detection and analysis of γ-H2AX. Its main advantage is the high speed of analysis of γ-H2AX content in a cell population. Immunofluorescence microscopy technique is an alternative to flow cytometry. It takes much more time than flow cytometry, but allows the quantitative analysis of the number of γ-H2AX foci, and determination of their spatial distribution within individual cell nuclei.

The number of γ-H2AX foci can be counted in cells at different phases of the cell cycle detected using antibodies to the proliferation marker Ki-67, and by the incorporation of 5-ethynyl-2′-deoxyuridine (EdU) which labels cells during replication.

Thymidine analog EdU is rapidly incorporated into cellular DNA during DNA replication. After cell fixation, it can be covalently cross-linked with fluorescent azide by a "click" chemistry reaction. It identifies S-phase cells in an asynchronous cell population. The advantages of EdU incorporation over 5-bromo-2′-deoxyuridine (BrdU) are described elsewhere [9].

Ki-67 is a proliferation marker that has characteristic staining patterns at different stages of the cell cycle, as demonstrated in fibroblasts and embryonic stem cells [10–13]. G0 cells are Ki-67 negative. Ki-67 distribution in S-phase and G1 is similar and is characterized by a low level of nucleoplasm staining and bright staining of nucleoli. S-phase cells can be discriminated by EdU incorporation, and G2 cells can be easily recognized by intensive staining of round-shaped nucleoli (Fig. 1).

Here, we describe the optimized procedure for the detection of γ-H2AX foci in mammalian fibroblasts grown on glass coverslips. In this procedure, the cells are fixed and stained directly on the coverslips. A similar approach uses in-solution labeling of γ-H2AX in mononuclear cells prior to their fixation on microscope slides [14].

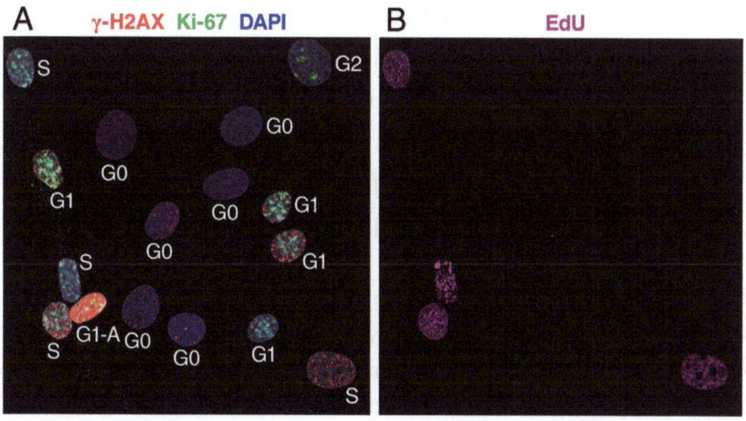

Fig. 1 Visualization of γ-H2AX foci in Syrian hamster fibroblasts in different phases of the cell cycle after the action of the radiomimetic drug bleomycin. After 30 min incubation with bleomycin (final concentration 50 μg/mL) and 10 μM EdU, cells were fixed, incorporated EdU was crosslinked with Alexa Fluor 647-azide, and then cells were immunostained with anti-γ-H2AX and anti-Ki-67 antibodies as described in the text. The images represent maximal projections of confocal sections collected using confocal microscope Leica TCS SP5. (**a**) γ-H2AX foci (*red*) induced in cells at different phases of the cell cycle (G1, S, G2) characterized by different Ki-67 staining patterns (*green*). G1-A—marks the cell homogenously stained with anti-γ-H2AX antibody that is characteristic for early stage of apoptosis. DNA is counterstained with DAPI (*blue*). (**b**) S-phase cells shown in "**a**" can be recognized by Alexa Fluor 647 fluorescence (*magenta*)

2 Materials

2.1 Isolation of Fibroblasts from Animal Tissues

1. Cell culture mediums: DMEM with low glucose content or MEM (autoclavable) supplemented with 13–15% fetal calf serum (FCS), 2 mM glutamine, 100 U/mL penicillin, 100 μg/mL streptomycin (Invitrogen).
2. Dulbecco's phosphate-buffered saline containing calcium and magnesium (DPBS + Ca/Mg) for isolation of cells from tissues (ThermoFisher Scientific).
3. PBS—phosphate-buffered saline without calcium and magnesium obtained from any manufacturer or prepared by dissolving the reagents in ddH_2O from powder using the standard formulation (1× PBS: 137 mM NaCl; 2.7 mM KCl; 10 mM Na_2HPO_4;1.8 mM KH_2PO_4).
4. CO_2 incubator.
5. Enzymatic cocktail for tissue digestion: 10 mg/mL collagenase IV from Clostridium histoliticum, 2.5 U/mL dispase, 0.0001% Trypsin-EDTA in DPBS + Ca/Mg. Sterilize by filtering through Millipore filter with 0.2 μm pore size.
6. Tweezers and razors (sterilized by dry heating).

7. Thermo-Shaker.
8. Sterile polypropylene 50 mL tubes.
9. Centrifuge for 50 mL tubes.
10. Pipettes with 1000 μL, 100 μL, and 10 μL sterile tips.
11. Sterile 100 mm diameter Petri dishes.
12. Sterile flasks with surface area 25 cm^2.

2.2 Immunofluorescence Microscopy

1. Glass slides and 1 mm thick glass coverslips.
2. Humid chamber for incubation of cells with antibody solution. It can be prepared from 150 mm diameter Petri dish. Put two pieces of filter paper soaked in dH_2O near the side walls of the dish.
3. PBS (as indicated in Subheading 2.1, **item 3**).
4. PBS containing 0.1% Tween 20.
5. Parafilm M Laboratory film (Sigma-Aldrich).
6. 37% Formaldehyde solution (Sigma-Aldrich).
7. Blocking solutions: 1% and 0.5% Blocking Reagent (BlR) (Roche). Prepare by diluting 5% stock solution in PBS and adding 0.02% Tween 20. 1% BlR solution is used for blocking prior to incubation with antibodies, and 0.5% BlR solution is used as antibody dilution buffer (*see* **Notes 1** and **2**).
8. Antibodies: mouse monoclonal anti-phospho-histone H2AX antibody (Millipore), rabbit polyclonal antibodies to Ki-67 (Abcam), secondary goat anti-mouse Alexa Fluor 568-conjugated antibodies (Invitrogen), goat anti-rabbit Alexa Fluor 488-conjugated antibodies (Invitrogen).
9. 0.5% Triton X-100 in PBS.
10. Click-iT EdU Imaging Kit (Invitrogen), containing EdU (5-ethynyl-2′-deoxyuridine), Azide-Alexa Fluor 647, 100 mM aqueous $CuSO_4$ solution, buffer additive, etc.
11. 3% BSA in PBS.
12. DAPI: 4′,6-diamidino-2-phenylindole (DAPI) 0.5 μg/mL solution in PBS.
13. Antifade mounting media with or without DAPI. In the first case, no preliminary staining with DAPI is needed.
14. Laser scanning confocal microscope such as Leica TCS SP5 system equipped with HCX PL APO 100×/1.4 and 40×/1.25 oil immersion objectives, 488 nm argon, 543 nm HeNe and 405 nm diode lasers and Leica LAS AF software. Any other confocal microscope model with similar characteristics can be used.
15. IPLab (Scananalytics, Inc.) and Adobe Photoshop programs for analysis of images.

3 Method

The duration of procedure is around 4 h 45 min.

1. Isolate primary fibroblasts from newborn animal or embryo (*see* **Note 3**).
2. Grow primary fibroblasts on 18 × 18 mm or 22 × 22 mm glass coverslips placed in 35 mm diameter Petri dishes in MEM supplemented with 10% fetal calf serum (FCS), 2 mM L-glutamine, 100 units/mL penicillin, and 100 μg/mL streptomycin at 37 °C in a humidified atmosphere with 5% CO_2. For DSB induction, cells can be treated with bleomycin or X-ray irradiated.
3. 30 min before cell fixation, add EdU to the culture medium to the final concentration of 10 μM.
4. Rinse coverslips with cells two times in PBS before fixation.
5. Fix cells with 4% formaldehyde in PBS for 15 min at RT.
6. Rinse coverslips two times with 3% BSA solution in PBS.
7. Permeabilize cells with 0.5% Triton X-100 in PBS for 20 min at RT with shaking.
8. Rinse coverslips two times with 3% BSA solution in PBS.
9. Prepare Click-iT cocktail including Azide-Alexa Fluor 647 triethyl ammonium salt (*see* **Note 4**).
10. Drop 50 μL of cocktail for each coverslip on a piece of Parafilm on the bottom of a Petri dish. Lay coverslips over 50 μL drops of reaction mixture. Be sure that you put the coverslip on the drop of cocktail with its surface covered with cells down. Keep coverslips for 30 min in the darkness at RT.
11. Rinse coverslips two times with 3% BSA solution in PBS.
12. Incubate coverslips with cells 30 min in a Petri dish with 1% BlR/0.02%Tween 20 at 37 °C.
13. Centrifuge stock solutions of antibodies at 15,000 × *g* for 10 min at +4 °C to pellet protein aggregates.
14. Dilute antibodies in 0.5% BlR/0.02% Tween 20. Dilutions of primary antibodies: mouse monoclonal anti-phospho-histone H2AX antibody (1:100), rabbit polyclonal antibodies to Ki-67 (1:200). Dilutions of secondary antibodies: goat anti-mouse Alexa Fluor 568-conjugated antibodies (1:400), goat anti-rabbit Alexa Fluor 488-conjugated antibodies (1:400) (*see* **Note 5**).
15. For each coverslip, drop 50 μL of the mixture of primary antibodies on a piece of Parafilm placed on the bottom of a humid chamber. Incubate cells for 1 h at 37 °C.
16. After incubation, wash slides by shaking in PBS supplemented with 0.1% Tween 20 for 30 min. Change wash buffer three times.

17. For each coverslip, drop 50 μL of the mixture of secondary antibodies on a piece of Parafilm placed on the bottom of a humid chamber. Incubate for 40 min at 37 °C.
18. Wash cells as indicated in **step 16**.
19. Counterstain the cellular DNA with DAPI: place coverslips in 0.05 μg/mL DAPI staining solution in PBS for 10 min at RT.
20. Wash coverslips in PBS for 5 min and mount them on microscope slides in antifade mounting media (*see* **Note 6**).
21. Use confocal microscope for the analysis of γ-H2AX induction in different phases of the cell cycle in gamma-irradiated or drug-treated cells. Compare these data with the data obtained for untreated cells containing very few spontaneous γ-H2AX foci. Use the confocal microscope equipped with 405 nm diode, 488 nm argon, 543 nm HeNe, 633 nm HeNe lasers. Collect Z-stacks of confocal sections in 4 channels (Red, Green, Blue, and Magenta) and obtain maximum projections of sections in stacks. Export RGB images (Red—γ-H2AX, Green—Ki-67 and Blue—DAPI) and single-color Magenta image to TIFF for analysis.
22. Use the segmentation function in the IPLab program (Scananalytics, Inc.) for counting the number of γ-H2AX foci per nucleus and for measurement of the area occupied by each focus. First, convert Red channel-images of RGB images to gray scale in Photoshop and save them in the TIFF format without image compression. Select the cells at a specific phase of the cell cycle that you are interested in according to Ki-67 staining in Green channel image and Magenta image representing EdU incorporation. Open the Red channel image converted to gray scale in the IPLab program. Select <Segmentation> command in <Analyze> menu and segment/threshold the foci from background. Using <Set Measurements> command from <Analyze> menu, choose the measurements you are interested in (densitometry and shape parameters of foci).

    Select each nucleus needed for analysis as a region of interest (ROI) using the appropriate tool in the ROI-tool toolbar. Choose <Quantify Segments> command from <Analyze> menu. Press OK and the data table will be generated, where the number of foci, focus area, and densitometry parameters inside the ROI are calculated. Export the table to Microsoft Office Excel.
23. Sometimes, the colocalization of γ-H2AX foci and foci of other repair proteins of interest may be needed. It can be performed in non-S-phase or S-phase cells using EdU incorporation for the identification of S-phase (*see* **Note 7**).

4 Notes

1. 5% stock solution of Blocking Reagent (BlR) is prepared from a powder: 10 mL Tris–HCl pH 8.0 and approximately 70 mL ddH_2O are added to 5 g of the powder, then the mixture is heated to 60 °C with gentle shaking. After dissolving, make up the volume to 100 mL with ddH_2O. Stock solution is aliquoted and kept at −20 °C before use.
2. Alternatively, 1% BSA or 10% fetal calf serum could be used as a blocking solution.
3. Description of technique for enzymatic isolation of fibroblasts from embryos or newborn animals:

   All manipulations with animals were carried out in strict accordance with the recommendations in the Guide for the Care and Use of Laboratory Animals of the National Institutes of Health, Washington DC, 1996.

   Put a fragment of a newborn animal skin in sterile Petri dish containing 0.5 mL of culture medium without serum and separate a piece 1.5 cm × 1.5 cm in size. Using sterile razor and tweezers, dissect the piece into smaller pieces (1 mm × 1 mm in size, or less). Transfer the pieces to 50 mL sterile tube. Wash the pieces twice with PBS, centrifuge at 120 × *g* at RT. Add 3 volumes of enzymatic cocktail for tissue digestion and incubate 30 min at 37 °C in thermo-shaker. Dissociate cells by gentle pipetting using sequentially 25 mL, 10 mL, and 5 mL sterile serological pipettes gradually decreasing the inside tip diameter of pipettes. Dilute cell suspension in DPBS + Ca/Mg to bring the final volume up to 50 mL. Centrifuge cells at 270 × *g* for 5 min at RT. Resuspend cells with PBS and centrifuge as described above. Plate cells on Petri dishes 100 mm in diameter or in flasks with surface area 25 cm^2 in culture medium containing supplements, and place in a CO_2 incubator. The cells are at the first passage after you split growing cells in ratio 1:2. When a monolayer is formed, split the cells at a 1:2 ratio.
4. For example, for 2 coverslips, prepare 100 μL of the mixture of the components from the Kit: 86 μL 1× Click-iT reaction buffer working solution; 4 μL $CuSO_4$ solution (100 mM); 0.24 μL Alexa Fluor 647-azide working solution; 10 μL Reaction buffer additive prepared fresh by diluting 10× stock solution in deionized water.
5. Keep diluted antibodies at +4 °C before use.
6. Mounted samples can be stored for several months at +4 °C in plastic boxes.
7. In such a case, Ki-67 staining is replaced with immunofluorescence staining of the protein of interest and individual confocal

sections of the nuclei are collected. The degree of the overlap of foci can be estimated with the "Colocalization" plugin of the free graphical analysis software ImageJ (National Institutes of Health, Bethesda, MD). Earlier, we presented a detailed description of this approach for colocalization analysis of γ-H2AX foci and foci of several DSB repair proteins in irradiated Syrian hamster fibroblasts [13].

Acknowledgments

This work was supported by the Russian Science Foundation (Grant # 16-14-10240).

References

1. Rogakou EP, Pilch DR, Orr AH et al (1998) DNA double-stranded breaks induce histone H2AX phosphorylation on serine 139. J Biol Chem 273:5858–5868
2. Rothkamm K, Löbrich M (2003) Evidence for a lack of DNA double-strand break repair in human cells exposed to very low x-ray doses. Proc Natl Acad Sci U S A 100: 5057–5062
3. Turinetto V, Giachino C (2015) Multiple facets of histone variant H2AX: a DNA double-strand-break marker with several biological functions. Nucleic Acids Res 43:2489–2498
4. Povirk LF (2012) Processing of damaged DNA ends for double-strand break repair in mammalian cells. ISRN Mol Biol 2012:1–16
5. Seiler JA, Conti C, Syed A et al (2007) The intra-S-phase checkpoint affects both DNA replication initiation and elongation: single-cell and DNA fiber analyses. Mol Cell 27:5806–5818
6. Zhao H, Traganos F, Darzynkiewicz Z (2010) Kinetics of the UV-induced DNA damage response in relation to cell cycle phase. Correlation with DNA replication. Cytometry A 77:285–293
7. Staszewski O, Nikolova T, Kaina B (2008) Kinetics of gamma-H2AX focus formation upon treatment of cells with UV light and alkylating agents. Environ Mol Mutagen 49: 734–740
8. Ewald B, Sampath D, Plunkett W (2007) H2AX phosphorylation marks gemcitabine-induced stalled replication forks and their collapse upon S-phase checkpoint abrogation. Mol Cancer Ther 6:1239–1248
9. Mead TJ, Lefebvre V (2014) Proliferation assays (BrdU and EdU) on skeletal tissue sections. Methods Mol Biol 1130:233–243
10. Kill IR (1996) Localisation of the Ki-67 antigen within the nucleolus Evidence for a fibrillarin-deficient region of the dense fibrillar component. J Cell Sci 109:1253–1263
11. Ghule PN, Becker KA, Harper JW et al (2007) Cell cycle dependent phosphorylation and subnuclear organization of the histone gene regulator p220(NPAT) in human embryonic stem cells. J Cell Physiol 213:9–17
12. Solovjeva LV, Demin SJ, Pleskach NM et al (2012) Characterization of telomeric repeats in metaphase chromosomes and interphase nuclei of Syrian Hamster Fibroblasts. Mol Cytogenet 5:37
13. Solovjeva L, Firsanov D, Vasilishina A et al (2015) DNA double-strand break repair is impaired in presenescent Syrian hamster fibroblasts. BMC Mol Biol 16:18
14. Johansson P, Muslimovic A, Hultborn R et al (2011) In-solution staining and arraying method for the immunofluorescence detection of γH2AX foci optimized for clinical applications. Biotechniques 51:185–189

Chapter 18

RAPD-PCR as a Rapid Approach for the Evaluation of Genotoxin-Induced Damage to Bacterial DNA

Rosanna Tofalo and Aldo Corsetti

Abstract

RAPD PCR is a sensitive and reliable approach useful for the detection of DNA lesions due to environmental contaminants. In addition, this method is cost-effective, and can be performed in any laboratory having a DNA thermocycler and gel electrophoresis system. Here, we describe its application to identify genotoxin-induced DNA damage in foodborne bacteria. DNA alterations are detected through the analysis of electrophoresis profiles with the appearance or disappearance of new bands as compared to the non-mutated control. The described RAPD PCR procedure takes 6 h for completion. It uses small amounts of DNA and can reveal even low mutation rates.

Key words RAPD-PCR, Bacterial foodborne model, DNA damage, Mutation, Genotoxin

1 Introduction

Random amplification of polymorphic DNA (RAPD)-PCR technique involves the amplification of "anonymous" DNA sequences, using single short primer (8–12 nucleotides). This whole genome PCR sampling method acts by priming at arbitrary sites to obtain a specific complex profile. Differences in the obtained profiles are mainly due to the presence or absence of a priming site or due to the changed distances between priming sites.

Generally, the amplified regions represent the hypervariable noncoding sequences that are species-specific. Therefore, in food microbiology research, RAPD-PCR has been used for microbial species identification and to determine strain-specific fingerprints in order to differentiate strains belonging to different yeast/bacteria species [1, 2]. The strain discrimination stringency by RAPD-PCR depends on the primers used for the analysis. Generally, the primers are short (about 10 bp with 40–70% GC content) and the distance between priming regions is maintained below 3–4 kb.

The RAPD-PCR approach has some limitations, such as the difficulty to distinguish whether a DNA segment is amplified from

Vladimir V. Didenko (ed.), *Fast Detection of DNA Damage: Methods and Protocols*, Methods in Molecular Biology, vol. 1644, DOI 10.1007/978-1-4939-7187-9_18,

a locus that is heterozygous (a single copy) or homozygous (two copies). In addition, RAPD primers tend to mostly screen GC-rich regions because of their high GC content [3]. Furthermore due to the random nature of amplification, both nuclear and organelle DNA may be amplified during PCR [3].

In addition, the reproducibility of RAPD-PCR is strongly affected by the purity, quantity, and quality of DNA templates, variations in cations concentration in the PCR buffer and, specifically, of $MgCl_2$ concentration in the reaction mix. In general, higher concentrations of $MgCl_2$ reduce specificity of the polymerization reaction by decreasing the enzyme fidelity. However, the reproducibility of RAPD-PCR can be significantly increased by the use of the standard protocols for DNA purification, PCR amplification, and agarose gel electrophoresis.

RAPD-PCR also presents interesting advantages. For example, it does not require any previous knowledge of genome sequence. It also uses very low amounts of DNA and does not need any specialized and expensive equipment. Furthermore, the approach is very sensitive, permitting the detection of a wide variety of DNA damage types (e.g., DNA adducts, DNA breaks) as well as mutations (point mutations and large chromosomal rearrangements).

This technique is widely applied in different research areas (medicine, forensic science, genetically modified organisms, etc.). For instance, the RAPD method has been used in studies examining genetic diversity, pedigrees, the identification of cultivars, pest resistance genes, sex markers and in the construction of genetic maps [4]. In addition, the RAPD assay has also been applied to detect genetic instability in tumors and DNA alterations induced by toxic compounds in animals, bacteria, and plants [3, 4].

In the field of ecotoxicology the RAPD-identified changes include alterations in band intensity as well as gain/loss of RAPD bands (Fig. 1). These bands are lost or gained when point mutations, inversions, deletions, additions, or gross chromosomal rearrangements affect primer annealing. In its ecotoxicology application the RAPD-PCR technique has been proposed as a biomarker assay to detect the environmentally caused DNA damage and mutations in both prokaryotes and eukaryotes [5–8]. DNA samples affected by genotoxic compounds are characterized by the appearance or disappearance of bands in comparison to the control DNA samples [4, 5]. These alterations are mainly due to the formation of DNA adducts, the appearance of mutations, or DNA strand breaks [4, 5]. Sometimes the bands' intensity can change because of large rearrangements occurring when chemicals interact with genomic DNA inducing copy number differences [3]. To evaluate the differences between the affected and nonaffected samples, a variety of statistical methods are used. These employ the specific similarity coefficients, such as the Pearson correlation coefficient [9].

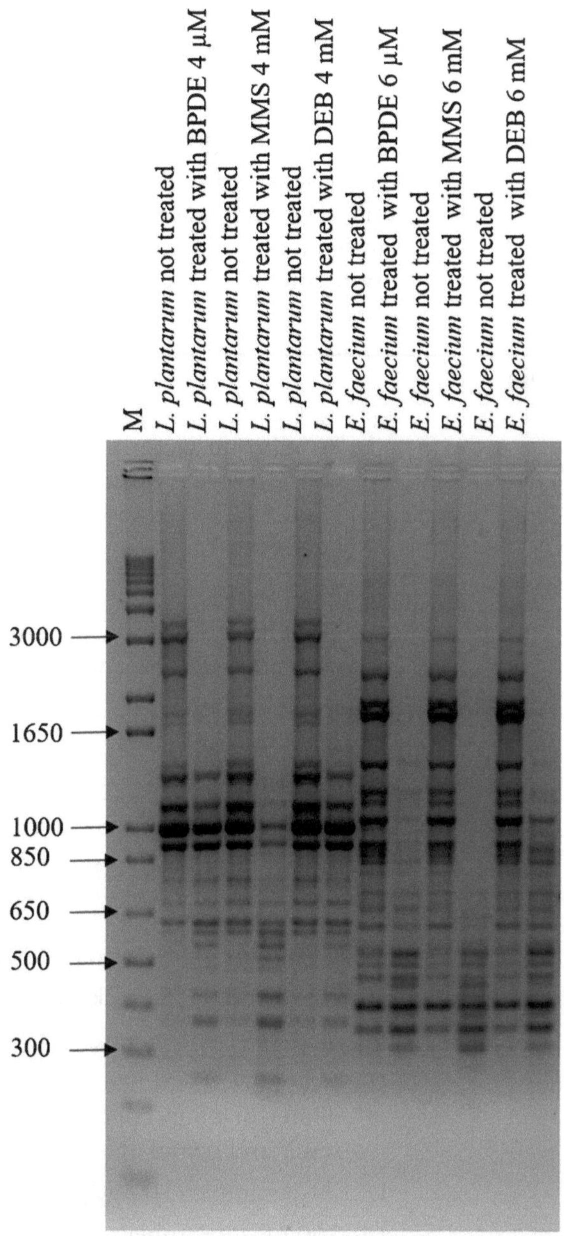

Fig. 1 M13 RAPD-PCR fingerprinting for *L. plantarum* ATCC 14917T and *E. faecium* DSMZ 20477T obtained for DNA samples untreated and treated with three different genotoxic compounds [benzo[a]pyrenediol epoxide (BPDE), methyl methanesulfonate (MMS) and 1,2,3,4-diepoxybutane (DEB)] used at different concentration. M: 1 kb DNA Plus ladder

Recently, Lanzone et al. [4] proposed the RAPD-based fingerprinting approach that uses the M13 primer to evaluate DNA damage and mutations induced by three genotoxic compounds forming covalent adducts with DNA [benzo[a]pyrenediol epoxide (BPDE), methyl methanesulfonate (MMS), and 1,2,3,4-diepoxybutane (DEB)] in foodborne bacteria (*Lactobacillus plantarum* ATCC 14917^T, *Enterococcus faecium* DSMZ 20477^T).

Here, we describe the eco-genotoxicology application of this M13-based RAPD fingerprinting approach. The technique uses *L. plantarum* and *E. faecium* as references for hazard identification in the environmental risk assessment. This method requires a small amount of DNA for the analysis and permits detecting the low mutation rates in approximately 6 h.

2 Materials

1. Foodborne bacteria: *E. faecium* DSMZ (Deutsche Sammlung von Mikroorganismen und Zellkulturen) 20477^T, *L. plantarum* ATCC (American Type Culture Collection) 14917^T (*see* **Note 1**).
2. MRS broth.
3. The Oxoid Anaerobic System (gas-generating pack with a jar) to cultivate anaerobic and microaerophilic organisms (*see* **Note 1**).
4. InstaGene™ Matrix (Biorad) for the preparation of PCR-quality DNA.
5. Centrifuge.
6. Spectrophotometer to determine DNA concentration.
7. 10× PCR buffer: 200 mM Tris–HCl, 500 mM KCl, pH 8.4.
8. 50 mM $MgCl_2$.
9. Ultra-pure sterile water (8 MΩ cm at 25 °C).
10. 2.5 mM dNTPs stock solution.
11. *Taq* DNA Polymerase (5 U/μL) (*see* **Note 2**).
12. Lyophilized primer M13 (5′-GAGGGTGGCGGTTCT-′3) diluted at a final concentration of 10 μM with ultra-pure sterile water and stored at −20 °C.
13. DNA sample (10 ng/μL).
14. 200 μL PCR tubes.
15. Thermocycler.
16. Agarose of electrophoretic grade.
17. DNA gel electrophoretic apparatus.
18. Power supply (220 V and 150 mA).
19. Gel electrophoresis 1× TAE buffer: 40 mM Tris-acetate, 1 mM EDTA, pH 8.0 (*see* **Note 3**).
20. Ethidium bromide: 0.5 μg/mL in 1× TAE buffer to visualize DNA.

21. 6× gel electrophoresis loading buffer: 30% glycerol, 0.25% bromophenol blue, distilled water to 10 mL.
22. DNA molecular weight marker (100 bp–12 kb).
23. UV transilluminator/gel documentation system.
24. Fingerprinting II Informatix software (Biorad).

3 Methods

3.1 DNA Extraction

1. Take 1 mL of bacterial culture, centrifuge at 16,000 × *g* for 5 min, and wash the pellet with ultra-pure water twice. The cells are now ready for DNA extraction.
2. To extract bacterial DNA, resuspend the pellet in 200 μL of InstaGene matrix and incubate at 56 °C for 15–30 min.
3. Vortex at high speed for 10 s.
4. Incubate the samples at 100 °C for 8 min.
5. Vortex at high speed for 10 s.
6. Centrifuge at 16,000 × *g* for 1 min.
7. Recover the supernatant and store at −20 °C.
8. Measure DNA absorbance in samples by using spectrophotometer and confirm the purity of DNA by calculating the 260–280 nm ratio. A ratio of ~1.8 is generally accepted as "pure" for DNA.
9. Determine DNA concentration by running the samples on 0.8% agarose gel in 1× TAE buffer and comparing bands intensity to a DNA marker. Alternatively, DNA concentration can be established by a spectrophotometer.

   DNA concentration is estimated by measuring the absorbance at 260 nm, considering that an $A_{260\,\text{nm}}$ of 1.0 = 50 μg/mL pure ds DNA:

$$\text{Concentration } (\mu\text{g/mL}) = A_{260} \text{ reading} \times \text{dilution factor} \times 50\ \mu\text{g/mL}.$$

   Total yield is obtained by multiplying the DNA concentration by the final total purified sample volume:

$$\text{DNA yield } (\mu\text{g}) = \text{DNA concentration} \times \text{total sample volume (mL)}.$$

3.2 RAPD-PCR Amplification

1. Perform RAPD-PCR analysis using M13 primer. Prepare PCR reaction mixture on ice, containing:

   2.5 μL—10× PCR buffer (Tris–HCl 20 mM, KCl 50 mM, pH 8.4).

   μL—$MgCl_2$ (50 mM)—3 mM final concentration.

   2.0 μL—dNTPs (2.5 mM dNTPs stock solution).

2.5 μL—M13 primer (1 μM final concentration).

0.4 μL—Taq (2 U/μL final concentration).

15.1 μL—Ultra-pure sterile water.

μL—DNA template (10 ng/μL).

25 μL—Total.

Include a negative control.

2. Perform amplification with an initial denaturation step at 94 °C for 4 min followed by 35 cycles consisting of 30 s at 94 °C, 20 s at 45 °C, 2 min at 72 °C, and a final extension of 7 min at 72 °C.

   Determine the repeatability of RAPD-PCR fingerprints by triplicate loading of independent replicate reaction mixtures prepared with the same strain [10].

3.3 Agarose Gel Electrophoresis

1. Prepare 1.5% (w/v) agarose gel in 1× TAE-buffer (*see* **Note 4**).
2. Mix PCR products with 1.5 μL of the loading buffer and load a molecular weight ladder into the first and last lanes of the gel and the other samples into the additional wells of the gel (*see* **Note 5**).
3. Run the gel in 1× TAE-buffer for 30 min at 80 V and then for 2 h at 100 V. At the end of the run switch off the power supply and transfer the gel to the ethidium bromide staining solution (*see* **Note 6**).
4. Let the gel stain for 30 min in the dark, then wash with distilled water for 15 min. Place the gel under an UV transilluminator, and take a photograph.
5. Calculate similarities among profiles using the Pearson coefficient [4, 9]. Perform clustering analysis by means of the UPGMA (Unweighted Pair Group Method with Arithmetic Mean) method using Fingerprinting II Informatix software.

4 Notes

1. Store bacteria (*E. faecium* and *L. plantarum*) at −80 °C in the MRS broth containing 20% (v/v) glycerol. Before experimental use, propagate bacteria in MRS broth at 30 °C in microaerophilic conditions by using an Oxoid Anaerobic Gas Generating Kit.
2. Repeated freezing and thawing may cause degradation of PCR reagents (Taq DNA polymerase enzyme, diluted primers, and dNTPs solution). Prepare several aliquots and store at −20 °C.
3. Prepare 50× TAE buffer: 242 g of Tris base, 57.1 mL of glacial acetic acid, and 100 mL of 0.5 M EDTA, pH 8.0. Mix all

ingredients and make it to 1 L with deionized water. Place 20 mL of 50× TAE in 980 mL distilled water to obtain 1× TAE working solution.

4. Try to avoid bubbles, which could alter gel structure and let the agarose solidify (takes about 30 min at room temperature). Any bubbles can be pushed away from the well comb or toward the sides/edges of the gel with a pipette tip.
5. It is important to use the same DNA molecular weight marker for all experiments as it permits reliable comparisons of the profiles obtained in different experiments. This also simplifies the preparation of the databases containing all experimental data.
6. Ethidium bromide is a potential carcinogen. Wear gloves, safety glasses, and lab coat when you work with it.

References

1. Ivey ML, Phister TG (2011) Detection and identification of microorganisms in wine: a review of molecular techniques. J Ind Microbiol Biotechnol 38:1619–1634
2. Tofalo R, Perpetuini G, Schirone M et al (2013) Biogeographical characterisation of *Saccharomyces cerevisiae* wine yeast by molecular methods. Front Microbiol 4:166
3. Atienzar FA, Evenden AJ, Jha AN et al (2002) Use of the random amplified polymorphic DNA (RAPD) for the detection of DNA damage and mutations: possible implication of confounding factors. Biomarkers 7: 94–101
4. Lanzone V, Tofalo R, Fasoli G et al (2016) Food borne bacterial models for detection of benzo[a] pyrene-DNA adducts formation using RAPD-PCR. Microb Biotechnol 9: 400–407
5. Atienzar FA, Jha AN (2006) The random amplified polymorphic DNA (RAPD) assay and related techniques applied to genotoxicity and carcinogenesis studies: a critical review. Mutat Res 613:76–102
6. Liu W, Yang YS, Li PJ et al (2009a) Risk assessment of cadmium contaminated soil on plant DNA damage using RAPD and physiological indices. J Hazard Mater 161:878–883
7. Liu W, Zhou QX, Li PJ et al (2009b) DNA mismatch repair related gene expression as potential biomarkers to assess cadmium exposure in *Arabidopsis* seedlings. J Hazard Mater 167:1007–1013
8. Wong CKC, Yeung HY, Cheung RYH et al (2000) Ecotoxicological assessment of persistent organic and heavy metal contamination in Hong Kong coastal sediment. Arch Environ Contam Toxicol 38:486–493
9. Pearson K (1926) On the coefficient of racial likeliness. Biometrika 18:105–117
10. Tofalo R, Schirone M, Telera GC et al (2011) Influence of organic viticulture on non-*Saccharomyces* wine yeast populations. Ann Microbiol 61:57–66

Chapter 19

Rapid Detection of γ-H2Av Foci in Ex Vivo MMS-Treated Drosophila Imaginal Discs

Varandt Y. Khodaverdian and Mitch McVey

Abstract

In *Drosophila melanogaster*, DNA double-strand breaks (DSBs) created by exposure to gamma or X-ray radiation can be quantified by immunofluorescent detection of phosphorylated histone H2Av (γ-H2Av) foci in imaginal disc tissues. This technique has been less useful for studying DSBs in imaginal discs exposed to DSB-inducing chemicals, since standard protocols require raising larvae in food treated with liquid chemical suspensions. These protocols typically take 3–4 days to complete and result in heterogeneous responses that do not provide information about the kinetics of DSB formation and repair. Here, we describe a novel and rapid method to quantify DSBs in imaginal discs cultured ex vivo with methyl methanesulfonate (MMS) or other DSB-inducing chemicals. The described method requires less than 24 h and provides precise control over MMS concentration and exposure time, enabling reproducible detection of transient DSBs. Furthermore, this technique can be used for nearly any chemical treatment and can be modified and adapted for several different experimental setups and downstream molecular analyses.

Key words Drosophila, Imaginal disc, Mutagen, Cell cycle, DNA damage response, Sensitivity assay, Apoptosis

1 Introduction

DNA damage tolerance is the process by which replication forks bypass DNA lesions during S-phase through the use of specialized translesion synthesis (TLS) polymerases and or homologous recombination [1]. Mutants defective in DNA damage tolerance are susceptible to prolonged replication fork stalls and fork collapse following exposure to DNA-damaging agents, resulting in DNA double-strand breaks (DSBs) and genome instability [2]. To study damage tolerance, organisms can be treated with a number of mutagens, one of which is the DNA alkylating agent, methyl methanesulfonate (MMS). While MMS does not directly cause DNA DSBs, the resulting 7-methylguanine and 3-methyladenine DNA lesions can stall and collapse replication forks in the

Vladimir V. Didenko (ed.), *Fast Detection of DNA Damage: Methods and Protocols*, Methods in Molecular Biology, vol. 1644, DOI 10.1007/978-1-4939-7187-9_19,

absence of functional damage tolerance responses, generating DSBs and eventually leading to cell death [3–6].

In *Drosophila melanogaster*, various genotoxic chemicals have been used to identify proteins that function in the cellular responses to DNA damage [7, 8]. Many of the pathways and proteins that act in these responses are conserved between Drosophila and humans, making Drosophila an attractive model organism to study metazoan damage tolerance and repair in the context of a whole organism. Well-established mutagen sensitivity assays that involve treatment of larvae with alkylating or crosslinking agents have been used to elucidate responses to various DNA-damaging agents [9]. However, these sensitivity assays utilize adult viability as a quantifiable endpoint and take an average of 20 days. Furthermore, they cannot elucidate the intermediate molecular effects of DNA damage.

To characterize the molecular responses to DNA damage in Drosophila, as well as to determine the underlying causes of lethality, investigators frequently utilize third instar larval imaginal discs. Imaginal discs are tissues consisting of diploid precursor cells fated to develop into adult appendages during metamorphosis; the growth and patterning of which are highly regulated [10]. Patterned early in embryogenesis, imaginal discs of third instar larvae are relatively large, easily accessible by dissection, and thin, making them highly suitable for fixing and immunostaining. Immunofluorescent studies of imaginal discs provide spatial resolution of protein modification and localization within cells and the tissue as a whole [11].

Apoptosis, checkpoint activation, cell cycle progression, protein foci, and DNA DSBs have all been studied in imaginal discs of larvae exposed to γ-ray irradiation (IR) [12–14]. These studies are possible because IR can penetrate the larval cuticle, providing accurate control of exposure and dosage. Unfortunately, imaginal disc studies have not been readily adapted for other treatments. This is largely because most techniques used to treat larvae with DNA-damaging agents require larval consumption of mutagen-dosed food over several hours to days. While this method of drug treatment is suitable for sensitivity assays, it does not provide the fine temporal control necessary to observe transient cellular responses to DNA damage prior to repair or apoptosis.

Here, we describe a method to rapidly culture and chemically treat third instar imaginal discs ex vivo for downstream immunofluorescent detection of DSBs. Compared to currently published protocols, which involve treating early-stage larvae and then quantifying DSBs in dissected imaginal discs 3–4 days later, our method takes less than 24 h. Partially dissected third instar larvae imaginal discs are cultured for 5 h in a simple culture media [15], with and without MMS or other DSB-inducing chemicals. They are then fixed, probed for γ-H2Av (phosphorylated histone H2Av, which is analogous to γ-H2AX in vertebrates and serves as a marker of DSBs) [16], and imaged using fluorescence. During the 5 h incubation period, approximately 90% of all wing imaginal disc cells should have entered S-phase [17], allowing for DSB formation in most genetic backgrounds. This technique allows for accurate temporal exposure to precise concentrations of any chemical, enabling detection of transient cell signals, protein modifications, and foci formation in imaginal discs. Possible applications of the technique include the study of checkpoint activation (Chk1 phosphorylation), cell division (histone H3 phosphorylation), apoptosis (cleaved Caspase-3), and DNA repair (RAD51 focus formation).

2 Materials

Reagents and materials listed below provide us with the results described here. Use of alternative reagents and materials may still work; however, modifications to the method may be necessary.

2.1 Larvae Imaginal Disc Dissection and Ex Vivo MMS Treatment

1. Dissection forceps, tip size D (#5 of biology) × 2.
2. Glass dissection dish or Sylgard® dish.
3. Dissecting microscope.
4. Petri dish.
5. 1.5 mL tube.
6. 96-well plate (flat-bottom).
7. 0.7% NaCl.
8. Culture media: 0.7% NaCl, 20% fetal bovine serum (FBS), 0.1% dimethyl sulfoxide (DMSO) [15].
9. Methane methylsulfonate (MMS), diluted to appropriate concentration in water and mixed well.

2.2 Wing Imaginal Disc Fixing and Antibody Staining

10. 37% formaldehyde.
11. Phosphate-buffered saline (PBS, 1×): 137 mM NaCl, 2.7 mM KCl, 10 mM Na_2HPO_4, 1.8 mM KH_2PO_4, pH 7.4.
12. PBS containing 0.1% Tween 20 (PBST).
13. PBS containing 0.3% Triton X-100 (PBSTx).
14. Blocking solution: 5% bovine serum albumin in PBSTx.
15. Primary antibody solution: 1:500 anti-γH2Av antibody (Rockland Inc.) in blocking solution.
16. Secondary antibody solution: 1:1000 goat anti-Rabbit IgG Rhodamine Red conjugated (Invitrogen), 500 μg/mL DAPI in blocking solution.
17. VECTASHEILD® Mounting Media (Vector Laboratories) (*see* **Note 1**).

18. Fluorescent microscope with Z-stack function.
19. Fiji image processing program with extended depth of field plugin [18, 19].

3 Methods

Culture, incubation, and wash times as well as reagent concentrations presented here have been extensively tested in our laboratory, and provide reproducible results if followed closely. Alternative incubation times and reagent concentrations can be adjusted to fit specific experimental demands, but have not been tested by our laboratory. For best results, use antibodies of high quality and purity.

3.1 Third Instar Larvae Imaginal Disc Dissection and Ex Vivo MMS Treatment

1. Collect third instar wandering larvae and place them in a small petri dish with 0.7% NaCl (*see* **Note 2**).
2. Transfer larvae one at a time to a glass-dissecting dish (or Sylgard® dish) containing 0.7% NaCl for dissection. In this protocol, "forceps 1" will designate the forceps held in the dominant hand, and forceps 2 will designate the forceps held in the nondominant hand.
   (a) Grasp the larvae two thirds (2/3) of the way from the head with forceps 1, and with forceps 2 grab the last one third (1/3) of the body (posterior) firmly, and rip it off. Place the posterior portion in a separate part of the dissecting dish to decrease debris.
   (b) Take the 2/3 anterior portion left behind and slip the cuticle, inside out, over the larva's head. The easiest way to do this is to place forceps 1 (closed) at the head of the larva and use forceps 2 to roll the cuticle over the head onto forceps 1. Then use forceps 2 to slide the inside-out cuticle off of forceps 1.
   (c) Holding the posterior portion of the cuticle with forceps 2, carefully remove the gut and fat tissue with forceps 1.
   (d) Look for the wing imaginal discs (WID) (the large, flat tissues that look like the continent of Africa). If you can see at least one, tease the gut and excess tissue from the cuticle. You will be left with the head with the discs attached to a piece of cuticle (Fig. 1) (*see* **Notes 3** and **4**).
3. Transfer cuticles to a 96-well plate with 97.5 μL of culture media. Place up to 8 cuticles per well (*see* **Note 5**).
4. In a chemical hood add 2.5 μL of freshly prepared 0.1% MMS to the cuticles in the culture media with a final concentration of 0.0025% MMS (*see* **Note 6**).

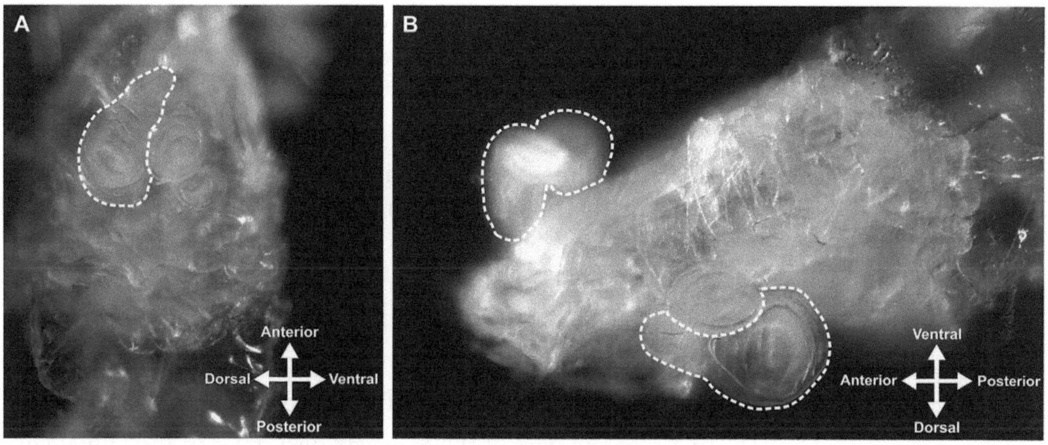

Fig. 1 Dissected cuticle of a third instar larva cleaned of all tissues except for the wing imaginal discs along with the adjacent leg and haltere discs. The wing imaginal discs are outlined with a dotted line. (**a**) A side view of the dissected cuticle presenting the right wing imaginal disc. (**b**) A top down view of the dissected cuticle with both wing imaginal discs visible

5. Incubate the cuticles for 5 h at 25 °C (the ideal temperature for Drosophila culture) with gentle shaking (35 rpm) protected from light (*see* **Notes 7** and **8**).

3.2 Imaginal Disc Fixing and Antibody Staining

6. Remove and dispose of the culture media containing MMS (*see* **Note 9**).
7. Rinse the cuticles by adding and subsequently removing 200 μL of cold PBST to the wells.
8. Add 200 μL of cold PBST to the wells and transfer the cuticles from the 96-well plate to 1.7 mL tubes. At this point you may pool together multiple wells of cuticles of the same treatment and discernable genotypes (*see* **Note 10**).
9. Remove the PBST transferred with the cuticles, and add 480 μL of cold PBST to the cuticles (*see* **Note 11**).
10. Fix the tissues by adding 20 μL of 37% formaldehyde to the PBST (final concentration is 1.48%) in the fume hood. Incubate for 30 min at room temperature while rocking on a nutating platform (*see* **Note 12**).
11. Centrifuge the samples for 10 s at 3000 × *g* to collect the cuticles and solution at the bottom of the tube. Remove formaldehyde mixture and dispose of it appropriately.
12. Wash four times for 5 min with 500 μL PBSTx at room temperature with rocking.
13. Block 30 min in 500 μL blocking solution at room temperature with rocking.
14. Centrifuge for 10 s at 3000 × *g* and remove blocking solution.

15. Add 500 μL of anti-γH2Av primary antibody solution and incubate overnight at 4 °C with rocking (*see* **Note 13**).
16. Centrifuge for 10 s at 3000 × *g* and remove primary antibody solution.
17. Wash four times for 5 min with 500 μL PBST at RT with rocking. On the final wash centrifuge for 10 s at 3000 × *g* and remove the PBST.
18. Add 500 μL of the secondary antibody solution and incubate for 2 h at RT, with rocking, protected from light.
19. Wash four times for 5 min with 500 μL PBST.
20. Transfer cuticles from the PBST onto a Sylgard® dish.
21. In a separate part of the Sylgard® dish, take an individual cuticle from the PBST droplet and place into a droplet of PBS. Using forceps, tease apart the WIDs from the rest of the cuticle and place the WID into a 30 μL VECTASHEILD® droplet on a microscope slide. Repeat for the rest of the cuticles (*see* **Note 14**).
22. Gently place a coverslip on the top of the VECTASHEILD® allowing for the VECTASHEILD® to spread completely under the coverslip. Seal the edges with clear nail polish and allow 10–15 min for the nail polish to dry protected from light (*see* **Note 15**).
23. Analyze samples for γ-H2Av foci at 10–20× magnification using a Z-stacking microscope and with filter sets compatible with DAPI and Rhodamine. γ-H2Av foci will appear bright red and fill the entire nucleus (Fig. 2) (*see* **Note 16**).

4 Notes

1. If another mounting media is used, ensure that it is compatible with DAPI and the fluorophore conjugated to the secondary antibody.
2. A slightly wet wooden probe works well to remove the larvae from the sides of bottles and vials. These larvae can be obtained from any source, such as bottles, vials, and grape plates. Heterozygotes and homozygotes can be distinguished by using a fluorescently marked balancer chromosome, such as one marked with a green fluorescent protein.
3. It is important to remove as much excess tissue as possible to ensure that the wing discs are not obstructed to allow uniform exposure of the tissues to the treatment. Removing the mandible and brain helps to decrease obstruction. Also removing the anterior spiracles (hand-like structures near the head) helps to expose the wing discs and prevents the cuticle from

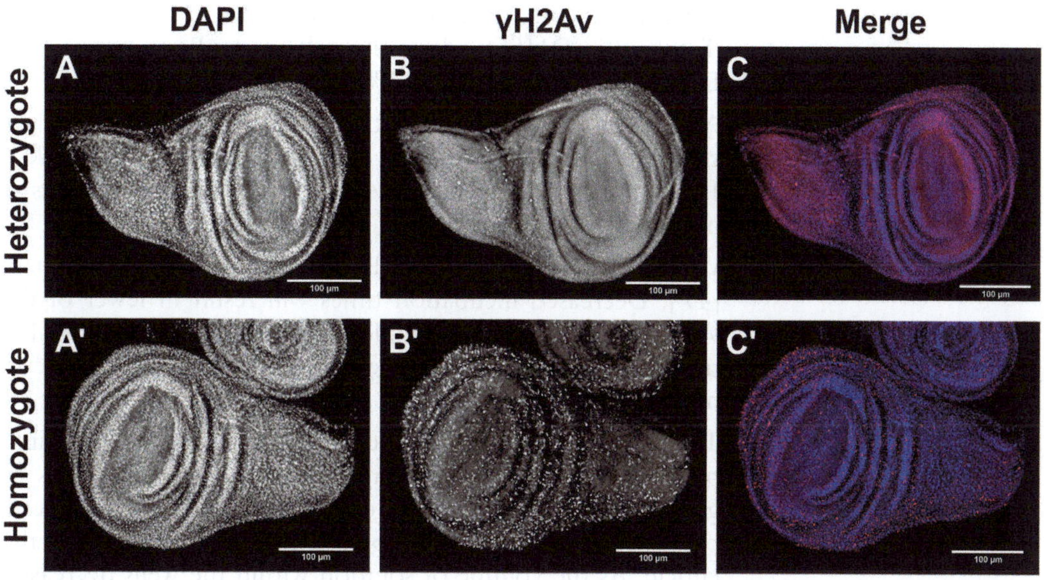

Fig. 2 Representative immunofluorescent results of wing imaginal discs from a MMS-sensitive mutant background. Heterozygotes (**a, b, c**) and homozygotes (**a′, b′, c′**) were treated, fixed, and stained as described in this protocol. The discs were visualized at 20× magnification. Z-stack images were deconvolved using the Zen 2 (Zeiss) program, then a composite image was generated through extended depth of field analysis. (**a, a′**) Nuclei are visible through DAPI staining. (**b**) A disperse fluorescent background is observed in heterozygous mutant discs, with a small number of γ-H2Av foci present. (**b′**) Distinct γ-H2Av foci are prevalent above background in homozygous MMS-sensitive mutants. (**c, c′**) A merge of both channels shows the clear presence of foci in the homozygous mutant compared to the heterozygote

scrunching up. Because the fine dissection results in less tissue to protect the WIDs, some discs may be lost during the procedure. Therefore, it is prudent to dissect a few more larvae than desired to ensure enough WID make it through the protocol.

4. WID are used in this protocol because they are the largest. Other imaginal discs can be used as well, however, culture time may need to be adjusted to correlate with the length of S-phase in these tissues [17].
5. Too many larvae per well will increase the probability that the cuticles may contact each other and obstruct WID exposure to culture media and treatment. 96-well plates are used to decrease the volume of MMS necessary. Larger incubation dishes and volumes may be utilized.
6. The concentration of MMS in the culture media is approximately 5–10 fold lower than what is necessary to observe lethality in a classic mutagen sensitivity assay. If larger volumes of chemicals are required, they should first be diluted in culture media to not drastically affect the concentrations of

the components within the culture media. If the chemical is dissolved in DMSO, adjust the culture media to have a final DMSO concentration of 0.1%.

7. As a safety precaution place the 96-well plate in a secondary container on top of bench paper on the orbital shaker. Incubation times can be adjusted for individual treatments as necessary.
8. Within 5 h 90% of WID cells should have entered S-phase [17]. Decreased incubation times will result in fewer of the WID cells entering and completing S-phase. Increased incubation times may result in increased apoptosis or a decrease in mitotically active cells.
9. To prevent the cuticle from being sucked up during pipetting, place the 96-well plate at angle allowing for the cuticles to gently settle toward one edge of the well. Using a p200 pipette begin pipetting the solution by placing the tip opposite to the cuticle. As the volume of solution within the wells decreases, gently move the tip along the edge of the well until you push the cuticles out of the way at the "bottom" of the well.
10. The easiest way to transfer the cuticles is to use a p1000 pipette and a pipette tip which has had the tip cut using a razor blade, creating a larger hole.
11. The best way to remove solutions without losing or harming the cuticles is to pipette using a p1000 pipette. Remove some of the solution, leaving about 100–200 μL within the tube. Gently pipette mix the remaining solution to lift the cuticles from the bottom, then place the pipette tip at the very bottom of the 1.7 mL tube and pipette the remainder of the solution.
12. Increasing the final formaldehyde concentration to 4% may allow for a decrease in fixing time.
13. The overnight incubation time of the primary antibody allows for complete diffusion of the antibody within the discs. To significantly shorten the protocol, an incubation for 90 min at room temperature may be used [11] or the antibody concentration can be increased.
14. Heterozygotes and homozygotes can be dissected and imaged on the same slide so long as there is a discernable fluorescent marker such as a balancer with GFP. Otherwise, genotypes should be treated, dissected, and imaged separately.
15. Slides can be kept at 4 °C for a few months, but the fluorescent signal may decrease over time. It is recommended to analyze the samples shortly after mounting.
16. To obtain the best images, discs should be imaged multiple times along the Z-axis (Z-stack), processed by deconvolution, and compressed into one image by extended depth of field algorithms.

References

1. Ghosal G, Chen J (2013) DNA damage tolerance: a double-edged sword guarding the genome. Trans Cancer Res 2:107–129. doi:10.3978/j.issn.2218-676X.2013.04.01
2. Branzei D, Foiani M (2010) Maintaining genome stability at the replication fork. Nat Rev Mol Cell Biol 11:208–219. doi:10.1038/nrm2852
3. Lundin C, North M, Erixon K et al (2005) Methyl methanesulfonate (MMS) produces heat-labile DNA damage but no detectable in vivo DNA double-strand breaks. Nucleic Acids Res 33:3799–3811. doi:10.1093/nar/gki681
4. Beranek DT (1990) Distribution of methyl and ethyl adducts following alkylation with monofunctional alkylating agents. Mutat Res 231:11–30
5. Tercero JA, Diffley JF (2001) Regulation of DNA replication fork progression through damaged DNA by the Mecl/Rad53 checkpoint. Nature 412:553–557. doi:10.1038/35087607
6. Fu D, Calvo JA, Samson LD (2012) Balancing repair and tolerance of DNA damage caused by alkylating agents. Nat Rev Cancer 12:104–120. doi:10.1038/nrc3185
7. Eeken JC, Romeijn RJ, de Jong AW, Pastink A, Lohman PH (2001) Isolation and genetic characterisation of the Drosophila homologue of (SCE)REV3, encoding the catalytic subunit of DNA polymerase zeta. Mutat Res 485:237–253. doi:10.1016/S0921-8777(01)00062-3
8. Ravi D, Wiles AM, Bhavani S, Ruan J, Leder P, Bishop AJR (2009) A network of conserved damage survival pathways revealed by a genomic RNAi screen. PLoS Genet 5:e1000527. doi:10.1371/journal.pgen.1000527
9. Henderson DS, Grigliatti TA (1992) A rapid somatic Genotoxicity assay in Drosophila-Melanogaster using multiple mutant mutagen-sensitive (Mus) strains. Mutagenesis 7:399–405. doi:10.1093/mutage/7.6.399
10. Cohen SM (1993) Imaginal disc development. In: Bate M, Martinez Arias A (eds) The development of *Drosophila melanogaster*, vol 2. Cold Spring Harbor Laboratory Press, Cold Spring Harbor, NY, pp 747–841
11. Klein T (2008) Immunolabeling of imaginal discs. Methods Mol Biol 420:253–263. doi:10.1007/978-1-59745-583-1_15
12. Brodsky MH, Sekelsky JJ, Tsang G et al (2000) mus304 encodes a novel DNA damage checkpoint protein required during Drosophila development. Genes Dev 14:666–678
13. Bilak A, Uyetake L, TT S (2014) Dying cells protect survivors from radiation-induced cell death in Drosophila. PLoS Genet 10:e1004220. doi:10.1371/journal.pgen.1004220
14. Wells BS, Johnston LA (2012) Maintenance of imaginal disc plasticity and regenerative potential in Drosophila by p53. Dev Biol 361:263–276. doi:10.1016/j.ydbio.2011.10.012
15. Gatti M, Santini G, Pimpinelli S, Olivieri G (1979) Lack of spontaneous sister chromatid exchanges in somatic cells of *Drosophila melanogaster*. Genetics 91:255–274
16. Madigan JP, Chotkowski HL, Glaser RL (2002) DNA double-strand break-induced phosphorylation of Drosophila histone variant H2Av helps prevent radiation-induced apoptosis. Nucleic Acids Res 30:3698–3705
17. Adler PN, MacQueen M (1984) Cell proliferation and DNA replication in the imaginal wing disc of *Drosophila melanogaster*. Dev Biol 103:28–37
18. Schindelin J, Arganda-Carreras I, Frise E et al (2012) Fiji: an open-source platform for biological-image analysis. Nat Methods 9:676–682. doi:10.1038/nmeth.2019
19. Forster B, Van De Ville D, Berent J, Sage D, Unser M (2004) Complex wavelets for extended depth-of-field: a new method for the fusion of multichannel microscopy images. Microsc Res Tech 65:33–42. doi:10.1002/jemt.20092

Index

A

Accelerated technique 129, 130
  definition 174
Alexa Fluor 488 130–136, 141, 143, 146, 148, 190, 191
Alkaline halo assay (AHA) 76
Alkaline-labile sites 151, 157
Alterations of DNA integrity 152
Anaerobic and microaerophilic organisms 198
Annexin V binding 70
Anonymous DNA sequences 195
Antibiogram 96
Antibiotic resistance 97, 101
Apoptosis
  changes associated with 23
  ex vitro 24–28, 30–32, 35–38, 43, 88, 205
  in vitro 24–26, 28, 29, 31, 33–36, 38, 43, 140
  in vivo 24–27, 32, 35–37, 63, 71
  in human tumor xenografts 44, 53, 54
  induced by cisplatinum 26, 28, 29, 34
  induced by photodynamic therapy 24, 26, 33, 37
  induction of 7, 140, 143, 160
  p53-dependent 29
  quantify the presence of 29
Apoptotic
  blunt-ended DNA breaks 4, 114
  bodies 110, 170
  cell clearance 3, 5
  DNA cleavage 79, 167
Aqueous slide mounting media 171

B

Bacteria 96–102, 184
Bacterial
  DNA 160
  foodborne model 198
  susceptibility 97, 98, 100, 101, 103
Biomarker 41, 124, 196
Biotin-dUTP 142, 145
Blunt-ended DNA breaks labeling 3–5, 116–117, 168, 174
Brain tissue 23–26, 28, 29, 31, 34, 37, 38
Breast cancer cells 44
Breast cancer response prediction 42, 43, 55, 56
5-Bromo-2''-deoxyuridine (BrdU) 140, 143–146, 188

C

Camptothecin 29, 46, 67, 106, 107, 110, 124–127, 144, 146, 188
Cancer response imaging 30, 38
Cell cycle
  alterations 13–15, 17, 19
Cell death 23–26, 28, 30, 31, 33, 35–38, 42, 43, 47, 48, 52, 55–57, 105, 204
Cell suspension 5, 66, 67, 84, 85, 99, 102, 107, 127, 133, 135, 145, 152, 153, 155, 193
Chromatin 13, 23, 33, 76, 84, 115, 123, 145, 170, 187
Chromosomal
  rearrangements 196
Clearance of cell death 3, 114, 115
Cleaved Caspase-3 205
Clinical ultrasound imaging 44
Comet assay
  advantages of 76, 182
  traditional 182
  twelve gel 181–186
Computer-aided theragnosis (CAT) 43, 50–54
Confocal microscope 125, 189, 190, 192
Cultured mammalian cells 105–110, 129–131, 133–137

D

Deep red anthraquinone 7 (DRAQ7) 105–110
Dexamethasone 115–117
4'',6-Diamidino-2-phenylindole (DAPI) 7, 13, 16, 17, 19, 69, 102, 115–117, 125, 127, 130–132, 134, 136, 137, 140, 141, 143, 146, 189, 190, 192, 205, 208, 209
Digoxigenin 141, 142, 145, 171, 174
Digoxigenin-11-dUTP 174
DNA adducts 160, 196
DNA breaks
  blunt-ended 4, 114, 167–171, 173, 174
  DNase I-and DNase II-types of 5
  double-strand 75, 151, 160
  with 3'OH 3
  5'OH DNA breaks 4
  with 3' overhangs 174
  single-strand 82, 90, 151, 152, 156, 160
DNA content 13–15, 17–20, 33, 84, 142, 144, 147, 152–155

Vladimir V. Didenko (ed.), *Fast Detection of DNA Damage: Methods and Protocols*, Methods in Molecular Biology, vol. 1644, DOI 10.1007/978-1-4939-7187-9, © Springer Science+Business Media LLC 2017

DNA damage 76
 detection 5, 23–24, 27, 32, 34–38, 160
 effects *in vivo* 43, 47, 56, 57
 induced 29, 46
 labeling 115–117, 127, 140, 144–146, 188
 response 187, 204
DNA fragmentation 15, 42, 96, 97, 101–103, 114, 148, 167
DNA gyrase 96
DNA integrity 151–155, 157, 159
DNA repair 124, 156, 159–161, 163–165, 182, 205
DNA single-strand break 152
DNA single-strand breaks (DNA-DSBs) 82, 83, 85, 90
DNase II 3–5
DNase II-type break 3, 5, 113–118
Double-strand breaks (DSBs) 78, 83, 85, 124, 125, 127, 129, 133, 139–147, 187, 188, 191, 194, 203–205
DSB-inducing chemicals 204
Dual-hairpin 4

E

Ecotoxicology 159, 196
Electrophoretic apoptotic ladder 169
Erythrocyte nuclei 16
Exosomes 170
Expedient markers of apoptotic cell 63, 71
Express detection of DNase II cleavage 4, 113
Express techniques
 definition v
Extracellular vesicle 170
Ex vivo 24–32, 35–39, 43, 47, 56, 71, 88, 140, 203–211

F

FAM and TAMRA 5, 6, 8
Fast detection
 definition 156
Fast micromethod 151–157
Fast-halo assay (FAH) 76–89
Fibroblasts 130, 187–194
Fingerprinting approach 197, 198
Flow cytometric detection 13–19
Flow cytometry 13–15, 17, 19, 28, 29, 33, 62, 67–70, 105, 109, 129–132, 136, 137, 140, 142–145, 188
Fluorescence resonance energy transfer (FRET)
 assay 4, 5, 62
 donor and acceptor 4, 5, 7, 8, 10
 efficiency 8
 nanoblinker 4–9
 oligo 5, 6, 8
 probes 4–8
 system 8
 ultra-fast technique 5
Fluorescent nucleic acid dyes 106
Fluorochrome-labeled inhibitors of caspases (FLICA) 62–71
Fluoroquinolones 96–98
Foodborne bacteria 198
Forensic science 196
Formalin-fixed tissue sections 113–118
Förster radius 8

G

Genetic diversity 196
Genomic integrity 159
Genotoxicity
 in unknown environmental mixtures 15
 rapid assessment of 13–15, 18
 detection of cell cycle alterations 13–21
Genotoxin-induced damage 196–198
Genotoxins 18, 19, 151, 152, 155, 157, 196, 197

H

γ-H2Av
 analogous to γ-H2AX 205
 antibody 205–206
 foci 204, 205, 208, 209
 marker of DSBs 205
γ-H2AX
 detection 124, 129–136
 foci 187–192, 194
 immunofluorescence labeling 124
HeLa cells 81, 84, 124–127, 152, 153
Hemocytes 15–19
Hemolymph probe 16, 17, 19
High-throughput genotoxicity assay 76

I

Image cytometry 140–148
Imaginal disc 203–210
Immunocytochemical detection 62, 123, 126–127
Immunofluorescence 124, 136
 analysis 190
 for the detection of γ-H2AX 191–192
In situ ligation (ISL) 169
 for blunt-ended DNA breaks
 for detection of fragmented DNA 167, 169
 labeling solution 172
 probes 168, 170
In vitro markers of apoptotic cells 61–71
Indicators of DNA ploidy 16
Integrity assay 152, 153
Irradiation 76, 130, 133, 137, 204
Isothermal
 amplification 169, 170
 cyclic elongation reaction 169

L

Labeling of phagocytosis ... 114, 116–117
Large-scale genotoxicity screening ... 76
Laser scanning cytometry (LSC) ... 63–67, 69, 70, 140–143, 145–147
Ligatable blunt end ... 169
Ligation-mediated polymerase chain reaction (LMPCR) ... 169
Live cultured cells ... 23–30, 33–37
Lysosomal DNase ... 3, 114
Lysosomes ... 113, 114

M

M13-based RAPD fingerprinting ... 198
Machine learning ... 43
Mammalian
  cells ... 105–110
  DNase II ... 113–117
  fibroblasts ... 187–194
Marine
  mussels ... 15
  sponge ... 152
Maximum mean discrepancy (MMD) ... 53
Microaerophilic conditions ... 200
Microgel ... 97, 99, 102
Micromax assay ... 97, 99–100
Microvesicles ... 170
Mitochondrial
  DNA ... 106, 159
  maintenance ... 159
  potential ... 63
Mitotic arrest/catastrophe
  induced by radiotherapy ... 28, 29, 35
MMS-treated Drosophila imaginal discs ... 203–210
Multicolor flow cytometry ... 105
Mussel gills ... 152, 155–157
Mutagen ... 88, 204, 209
Mutagens ... 14, 15, 198
Mutations ... 96, 196, 197
Mytilus galloprovincialis ... 15

N

Nanoblinker ... 4–9
Necrosis
  in cultured U87 cells ... 7
  induction of ... 7
  verification ... 7
Nucleoids
  with high level of damage ... 100
  with low level of damage ... 100

O

3'OH groups ... 117, 167
5'OH DNA breaks labeling ... 4, 9
5'OH DNA probes ... 4, 115, 167
Oligoprobes ... 114, 117

P

Paraffin-embedded tissue blocks ... 116, 170
PCR-labeled DNA fragments ... 169, 171
PEG ... 168, 173
Phagocytic
  cells ... 113–115
  clearance ... 3–5, 7–10
  degradation of DNA from apoptotic nuclei ... 113
  digestion of DNA ... 3, 4, 114, 115
  reaction ... 9, 114
Phagocytosis
  assessment ... 7–8
  FRET assay for ... 7
  in suspensions ... 4, 5, 84
  in tissue sections ... 114, 115
  induction of ... 7
  marker ... 4
Phagolysosomes ... 3, 9, 113–115
Phosphatidylserine (PS) ... 63, 70
Phosphorylated H2AX ... 124, 126
PicoGreen ... 151–153, 156
Plasma membrane integrity ... 63, 68
PO-PRO-1 ... 105–108, 110
5' Phosphorylated ends ... 169, 173
5'PO4 DNA breaks ... 170
Pretreatment ... 8
Programmed cell death ... 23, 129
  (*see also* Apoptosis) 61
Peripheral blood mononuclear cells ... 4, 152
Propidium iodide (PI) ... 33, 63, 102, 140
Proteinase K ... 116–118, 171, 172, 174, 175

Q

Quantification of DNA damage ... 159–165
Quantitative
  assessment of cell viability ... 105
  PCR (qPCR) ... 160–165
  ultrasound ... 23–26, 38, 42, 43, 50, 52, 55
Quick detection ... 114
Quinolone ... 95–103

R

Radial diffusion assays ... 76
Random amplification of polymorphic DNA (RAPD)-PCR ... 195–200
RAPD primers ... 196
Rapid
  apoptosis detection ... 167, 169
  assay definition ... v

assessment ... 13–15, 18
cytometric analysis ... 70
Real-time PCR ... 161
Repair in mitochondrial, nuclear, and bacterial genomes ... 159–165
Resistance to quinolones ... 95–97, 101, 103
Restriction site ... 171, 174

S

Saponin-based permeabilization ... 130
SCID mouse ... 46
Sensitivity assays ... 204
Signal enhancement ... 169, 173
Single-strand breaks ... 82, 90, 160
Sma I ... 171, 174
Small animal imaging ... 27, 43, 57
Solid tissues ... 152, 153, 155
Sonix RP system ... 44, 47
Spectroscopic detection ... 23–26, 57
Strain-specific fingerprint ... 195
Strand scission factors (SSFs) ... 154, 156
Strain-specific fingerprint ... 195
Suberites domuncula ... 152
SYTO 16 ... 105–110
SYTO dyes ... 71

T

T4 DNA ligase ... 168, 171, 172, 175
Texture analysis ... 43, 50, 51
Theragnosis ... 50
THP-1cells ... 106
Tissue equivalent phantom (TEP) ... 44, 45, 50
Tissue sections ... 3, 4, 167–170
Trout and chicken erythrocyte nuclei ... 16
Trypan blue ... 81, 84, 125, 126
TUNEL assay
for analysis of DNA damage ... 54
variants ... 140
Twelve gel comet assay ... 181–186

U

Ultrasound
backscatter signal ... 23, 25, 26, 28, 45
data collection ... 24, 31–34, 47–50, 55–56
imaging of apoptosis ... 37, 41, 42, 44–52, 54–58
spectroscopy ... 25–26
spectrum analysis ... 25, 31, 32, 55–56
UV-C light ... 152

V

V(D)J recombination ... 173
Vaccinia DNA topoisomerase I ... 114

W

whole genome PCR sampling ... 195

X

Xenografts ... 26, 29, 33, 44, 53, 54
Xylene ... 116, 170, 172

Y

YOYO dye ... 155

Z

Zebra tail amplification (ZTA) ... 169–175
ZTA enhancement ... 173
Zebra tails ... 169, 170, 173, 174

MIX
Papier aus verantwortungsvollen Quellen
Paper from responsible sources
FSC® C105338

If you have any concerns about our products,
you can contact us on
ProductSafety@springernature.com

In case Publisher is established outside the EU,
the EU authorized representative is:
Springer Nature Customer Service Center GmbH
Europaplatz 3, 69115 Heidelberg, Germany

Printed by Libri Plureos GmbH
in Hamburg, Germany